Refactor

The CTO's Guide to
Reclaim Your Time and
Multiply Your Output

ETIENNE DE BRUIN

CTO
SENTINEL

REFACTOR
The CTO's Guide to Reclaim Your Time and Multiply Your Output

FIRST EDITION

Published by CTO Sentinel

ISBN: 978-1-967830-03-9 Hardcover
978-1-967830-04-6 Paperback
978-1-967830-05-3 eBook

For more information about the book or to contact the author,
visit www.etiennex.com

To Brittany Cotton, my partner at 7CTOs and a constant muse in this work. Your insight and support have shaped so much of this journey.

And to every CTO who has ever lost direction or control of their time—this book is for you. I hope it reconnects you with your love for technology, building, and creating.

Table of Contents

The Breaking Point

The DNA of a chicken remains the same from egg to chicken. However, the effect of that DNA is different depending on the stage of growth of the chicken. The same genome that produces the fuzzy down of a chick will produce feathers in a chicken.

–Scott Graves, cofounder of CTO Sentinel

You have years of experience as Chief Technology Officer (CTO): you have climbed the ladder, steered through mergers, scaled your team, and managed every critical outage your company has thrown at you. You are now a master at firefighting: you or your team patching code at two in the morning, cramming back-to-back meetings into breakfast and bedtime, and juggling road maps that shift by the hour. Yet, with the artificial intelligence (AI) revolution disrupting everything you thought you knew about technology development, a nagging question keeps surfacing: *Is this really how I want to spend the next years of my professional life?*

If you are reading this, you already know that something is out of balance. You are not alone. Across the tech industry, high-performing CTOs, usually seasoned professionals who have spent decades honing their craft, find themselves pulled in different directions, drowning in urgent demands and reactive crises. Their calendars groan under the weight of status updates, emergency bug fixes, and leadership dynamics that feel more like occupational hazards than opportunities for fulfillment.

But what if there were another way? What if you could redesign your role from the ground up: define clear boundaries, reclaim your time, and lead with intention instead of exhaustion? *Refactor* offers exactly that. Here, you find a blueprint for transforming a job into a position where you can deliberately architect, align with your organization's mission, and focus on your well-being.

In software development, *refactoring* refers to "reorganizing code without altering its external behavior." Doing so will make it cleaner, more efficient, and easier to maintain. Your role as CTO is no different. Over time, added responsibilities, shifting company priorities, and cultural baggage can turn what was once a clear, strategic leadership position into unrealistic expectations, interruptions, and burnout.

This book treats your career like it should be treated. You will learn how to avoid redundant meetings, undefined decision paths, and productivity anti-patterns, replacing them with sustainable practices that put you back in the driver's seat. By the end, you will have a lean, generative role definition that adapts to change without breaking but also scales with your ambitions.

You should read this book if you are

- **an overloaded CTO** who spends more time troubleshooting than strategizing, and your team depends on you for every emergency.

- **experiencing burnout signs** such as missing vacations, skipping date nights, and dreading the Monday morning scramble to patch another vulnerability.
- **an aspiring leader** who loves technology but wants to lean into your people and business skills, stepping beyond the keyboard to shape long-term growth.
- **a seasoned veteran** who has seen the highs of successful launches and the lows of critical incidents. You know the stakes and want to sustain peak performance without sacrificing your personal life.

This book is about reclaiming your role not just as a technology leader but also as a person with limited time and infinite demands. You have likely spent years reacting to everything around you. And you know what? It might have worked for you in the early stages of your career, but it's not sustainable if you want to rekindle the love that got you into technology in the first place. *Refactor* is an invitation to pause, reassess, and redesign the way you show up as a CTO.

It offers a clear progression: first, clarifying your role so it serves both you and your company; second, building teams and systems that operate without constant oversight; and finally, creating the time and space for deep, meaningful work. You won't find rigid prescriptions or flashy productivity hacks. What you will find is a set of flexible frameworks that evolve with your context. The idea is to help you lead deliberately rather than reactively.

If you are ready to stop surviving and start shaping the role on your own terms, move on to the next page. The transformation unfolds one deliberate decision at a time.

The Refactor Assessment

Measure the Design of Your Role

Most CTO roles aren't designed.
They evolve by accumulation.

The Refactor Assessment shows you how your role is functioning today.

You'll see:

- How clearly your mandate is defined
- Whether you're operating as a multiplier or a bottleneck
- How much usable autonomy you truly have

At the end, you'll receive your **Autonomy Score:** a direct signal of how much redesign your role requires.

This isn't a productivity quiz.
It's a systems diagnostic.

If you want leverage instead of drag, start here.

Scan the QR code or visit:
https://ctorefactor.com/assessment

PART I

ARCHITECTING YOUR CTO ROLE

CHAPTER 1

Identity Crisis: Who is a CTO Today?

Until you define your CTO identity, others will define it for you.

Step into any leadership meeting and ask, "What should our CTO be focused on right now?" and you will quickly discover there is no one agreed-upon answer. The expectations are numerous, sometimes conflicting, and often shaped by urgency more than clarity. The title remains the same, *CTO*, but the definition changes from boardroom to hallway, from start-up to scale-up.

This book is not about finding the *definition* of a CTO. It is about helping you discover *your* relationship to the role. Before you can effectively lead, you need a strong internal compass and a clear understanding of the value you bring, the role you want to inhabit, and the leadership version that aligns with your strengths and values. Without that clarity, your role will default to what everyone else wants it to be—and that is where the breakdown starts for you.

The Fractured Role

No two CTOs are the same. If you asked 10 companies what they expect from their CTO, you would get 10 very different interpretations. One wants a deep technologist. Another wants a business strategist. Some want product ownership. Others want infrastructure, culture, security, team building, and sometimes all of it, and somehow, you are expected to keep up.

Christina Maslach's research at UC Berkeley offers a crucial lens for understanding why the CTO "expectation gap" is not just a semantic or organizational problem but also a burnout risk (Maslach & Leiter, 2022). Maslach's model finds that burnout tends to arise when there is a mismatch between the individual and the job environment in any of six areas: workload, control, rewards, community, fairness, and values.

In the context of the CTO role, which often stretches across product, infrastructure, culture, security, people, and budgeting, every one of those dimensions is under stress:

- overwhelming workload (too many priorities)
- limited control (demands from executives or investors with little autonomy)
- unclear or insufficient reward (recognition, support, or resources)
- fractured "community" (Engineering vs. business tension)
- questions over fairness (who gets support or whose vision prevails)
- potential conflicts of values (entrepreneurial visions vs. technical or ethical judgment)

This alignment pattern helps explain why many CTOs (especially in fast-growing or hyperdemanding start-ups) burn out or underperform: not because of their personal weaknesses but because the role itself is structurally ill-defined and overloaded. Maslach's research warns against viewing burnout as an individual problem needing self-care or resilience training. The key for the process to work is to "fix the job, not the person" (Maslach & Leiter, 2022).

For a CTO, that means clearly defining which slice of the "corporate organization building" they own, negotiating realistic boundaries, cultivating clarity and fairness within the organization, and ensuring a supportive community. On the other hand, if this is not done, the broken expectations may lead to inefficiency, exhaustion, cynicism, and disengagement.

This is the fractured reality we live in, and if you are not intentional about what your role is and your relationship to it, it will be *defined for you.* This will be done by your CEO, other executives, board members, product managers, investors, or even the vaguely disgruntled voices in your Engineering team.

Not because the role itself is unclear, but because each leader approaches it differently. The core function of a CTO is like a triangle: Understand the company's goals, own and manage the technology budget, and build the organization that delivers the technology. That triangle does not change. What does change is how each CTO brings that function to life: what they focus on, what they delegate, and where they bring their unique energy.

Through 7CTOs, the peer community I founded in 2013, and Coach EXE, my coaching practice for executive teams, I hear the same story again and again. Talented, ambitious people who started out passionate and excited about the role now find themselves quietly crumbling under the weight of misaligned expectations. What was once a dream job has turned into a reactive spiral—days filled with ambiguity, constant demands, and no clear sense of what success is supposed to look like.

But *why?* Because execution in the role of CTO is not static, it's contextual. Situational. Evolving. But here is the paradox: While your responsibilities may shift based on the company size, stage, or need, it is essential that your role remains constant in purpose. The CTO is the architect of an organization. You are the designer of the technology organization within a larger system. And that system is a *complex adaptive system,* a term coined by John Holland at the Santa Fe Institute. He describes complex adaptive systems as ones "that involve many components that adapt or learn as they interact" (Holland, 2006). This means that they are not a machine to control but rather an ecosystem that needs to be nurtured and cared for.

If you do not embrace this concept, the role will reduce you to a glorified ticket-taker or firefighter. But when you own your identity and intentionally design your activities around it, everything changes. This is when things shift, and results become more tangible: Results are the by-product of the shift toward coherence in this complex adaptive system. Great results become far more likely when you are aligned and operating with coherence.

The CTO Levels™

Let's get more concrete. A key part of defining your CTO identity is understanding the scope of your current reality and the level of complexity you are operating within. While we have established that the fundamental role of the CTO remains unchanged, your day-to-day engagement with that role shifts as your company grows.

The Cynefin framework (Snowden & Boone, 2007) is useful here: It separates work into simple, complicated, complex, chaotic, and disorder domains. Most CTO responsibilities, such as scaling an Engineering team, architecting systems under evolving constraints, managing Product–Engineering tensions, and navigating outages, sit squarely in the complex domain, where cause and effect only become clear in hindsight. In this environment, rigid plans or fixed playbooks will not save you; you will need to go through experimentation, learning loops, and rapid decision-making.

This complexity also shifts as your company grows. Early-stage environments often feel chaotic, demanding decisive action just to restore stability. As the company matures, problems move into the complicated domain (one that is solvable with expertise and analysis) before inevitably drifting back into complexity as scale introduces new unknowns. A CTO who understands which domain they are currently operating in can calibrate their leadership: when to explore, when to standardize, when to delegate, and when to intervene directly.

Your identity as a CTO is not a static label but an adaptive stance shaped by context, clarified by constraints, and grounded in how you manage the range of problems facing you. For this reason, the *CTO Levels™ Framework* was created to help you look through a simple lens that will help you self-assess based on two dimensions:

- **Team size:** How many people report to you (this includes full- and part-time employees and contractors)?
- **Tech budget:** This is your sphere of financial responsibility for spending on people (not including your operational expertise).

From these two inputs, it is possible to place yourself on a level scale from 0 to 10 on the CTO Levels™ grid (find the grid at ctolevels.com). However, this is not a hierarchy: It is a *complexity map.* This assessment is essential because a "Level 3" CTO and a "Level 8" CTO are solving completely different problems, even if their roles remain the same.

For example, Dropbox grew rapidly from a few dozen engineers to hundreds. According to engineering leadership accounts, the original monolith architecture that once enabled speed became the very thing that slowed them down. Complexity arrived not in one big moment but through countless small decisions: adding new features, onboarding new hires, creating processes to manage exceptions. The story is common across software-as-a-service (SaaS) companies—what worked brilliantly at 10 engineers breaks catastrophically at 100.

Here is what I always ask CTOs:

- How many people are there in total in your tech organization?
- What is your budget for your tech professionals?
- What business outcomes are you truly accountable for?

Answering these questions brings clarity. And with clarity comes ownership. Further along, in Chapter 5, you will look deeper into the CTO Levels™ Framework, allowing you to fully understand and later implement the concept.

The Expectation Gap

Here is the painful truth: *No one knows what a CTO does.* Or worse: *They think they do.* This is what I call the *expectation gap.*

In this silent tension, your CEO wants visionary alignment, your engineers want architecture decisions, your investors want shipping velocity, and your project manager just wants to unblock Jira tickets. Honestly? It's exhausting.

Different studies show that technology sector CTOs experience tenures of 2–3 years compared to the overall C-suite average of 4.9 years—evidence that misaligned expectations create unsustainable conditions (*Age and Tenure,* 2020). So many of the CTOs I work with are stuck in this tangle of misaligned demands. They are overfunctioning in some areas while underfunctioning in others, and they are constantly unsure where they are actually adding value. The first step out? Map the expectations. Here is a quick exercise:

1. Write down what your CEO, COO, and CPO think your job is.
2. Compare this to how your team sees your role.
3. Now, compare *both* to what you *actually want your role to be.*

What you will see as a result is usually an empowering realization: The expectation gap allows you to design what it is *you* want to deliver to the organization. Mapping expectations doesn't change your core role; it helps you clarify where misalignment in execution is making that role harder to fulfill.

The Mirror Exercise

Grab a notebook or open a new document on your computer. This will help you see what you are doing currently and what you want to achieve.

- **Step 1:** List your current responsibilities. These include the actual tasks, meetings, and decisions you own week to week. Be brutally honest, and include everything, even the tasks you'd rather not admit you own.

- **Step 2:** Within the list, circle the items that energize you. Put a line through the ones that drain you.
- **Step 3:** Think about the role that would ignite your zone of genius. What would your calendar look like? What would you stop doing? What would you protect?

This is not fluff. This is *role design,* a conscious refactoring of your time and talent toward high-leverage outcomes. If you are not deliberate about this, you will end up compensating and reacting to others' needs instead of architecting your own contribution.

Refactor in Action

The room was buzzing with ideas for the next quarter's road map, but one voice cut through it all.

"Let's check with Leo. He's basically our project manager anyway."

Laughter rippled around the table. Leo, the CTO, forced a smile, but inside, it felt like a gut punch. Project manager? After years of building the core architecture, hiring the first engineers, and scaling the product, that's what he had become?

On the drive home, he couldn't shake it. His calendar had no space left for thinking, designing, or even reading a line of code. It was back-to-back meetings, putting out fires, and answering Slack threads at midnight. Somewhere along the way, he had traded building systems for babysitting them.

That night, staring at the ceiling, he admitted the truth: He no longer recognized his own role. He wasn't leading. He wasn't creating. He was maintaining everyone else's expectations—and losing himself in the process.

The realization didn't spark an overnight fix. It started with a scribbled question in his notebook: *"How do I get back to the work I actually want to be doing?"* That single line became the thread he began to pull, slowly unraveling the tangle of obligations he'd inherited.

The Invitation

If you have ever felt the weight of trying to be everything to everyone, you are not alone. The good news is that your role is more designable than it seems. You don't need permission to clarify your identity. You just need the courage to question the defaults and the willingness to redraw the map.

At the heart of contemporary systems thinking, as articulated by Herbert A. Simon (1973), complexity is not a defect to be fixed but a reality to be managed. In his seminal work on "nearly-decomposable systems," Simon argues that many of the organizations and institutions we navigate (companies, teams, and technical stacks) work because they contain interacting subsystems that are loosely coupled enough to evolve, yet integrated enough to function as a whole. What makes such systems effective is not simplification at all costs but designing interfaces, boundaries, and relationships so that each part can operate semi-independently while still contributing to a coherent whole.

For a CTO, embracing the model means treating the organization (Engineering, Product, and Operations) as a living, adapting structure rather than a rigid hierarchy. It means recognizing that solving every problem top-down or expecting perfect coordination across every team at every moment is both unrealistic and counterproductive. Instead, clarity around which subsystems you own, which empower you to act, and how you enable information flow between them becomes far more valuable than trying to micromanage every detail. Redrawing the map—defining boundaries, clarifying responsibilities, designing feedback loops, and letting parts behave autonomously—helps you shape a system that can grow, adapt, and survive complexity.

You have more control over your role than you think. But clarity is not handed to you: It is earned through reflection, conversation, and courageous boundary-setting. Start by being honest about your identity:

- What is your real role?
- What does your organization need from you?
- What do you want to bring to the organization?
- What will you no longer accept as your normal?

Being a CTO should not be a sentence. It should be a system that you get to design. Let this chapter be the beginning of that process. You are not just a CTO by title: You are also a system architect, a leadership designer, and a multiplier of talent and clarity. However, this will only happen if you claim that identity for yourself.

REFLECTION PROMPT

What part of your current CTO role was designed with intention, and what part was inherited by accident? Write it down. And then write down what needs to change.

Redefining your CTO identity is the first act of leadership in your organization and, most importantly, in your own life. When you begin to see yourself not as a passive participant but as the designer of your role, everything starts to change. The ambiguity becomes design space. The frustration becomes fuel. And your leadership begins to align not just with what the business demands but also with what you are uniquely wired to create.

However, it is important to keep in mind that clarity without design still leads to drift. Now that you have started naming who you are (and who you are not), it is time to build a role that works for you, and not against you. In the next chapter, we will start creating your CTO blueprint: a tactical and personal map that brings alignment between your energy, your value, and your company's needs. It's not about escaping responsibilities. It's about focusing on the ones that accelerate real progress, light you up, and create lasting impact.

CHAPTER 2

The Blueprint: Designing Your Authentic CTO Role

You can design a role that energizes you and creates real value—if you negotiate it with clarity and courage.

Many CTOs suffer in silence under the weight of a job description that keeps changing focus and priorities. Somewhere between scaling systems and surviving strategy shifts, the job became something they do *for others* rather than something they design *with intention*. Somewhere down the road, they adopted this mindset, and it hasn't changed since.

However, before we dive into designing your ideal role, I want to be clear about something foundational: *The CTO role is not subjective.* It is not a blank canvas or a choose-your-own-adventure leadership title. At its core, the job is fixed, with the task of aligning technology with business objectives, managing your tech budget with intention, and building the technology organization that delivers on those goals.

That is the function of the CTO. Where *Refactor* adds value is in the form: the way you personally engage with the function.

What you delegate. What you prioritize. That's the blueprint we will build here. Not to redefine the CTO role, but to help you bring it to life in a way that's sustainable, energizing, and high leverage.

Here is the truth: You don't seek the *right* to do your job; you already have it. What you need now is a *blueprint,* a structured, honest, and strategic way to articulate the CTO you want to be, the value you want to create, and the boundaries you will need to protect to stay in your zone of genius.

Let's build that blueprint together.

The CTO Role Canvas: Your Operating Manual

In my previous book, *CTO Excellence in 100 Days,* I introduced the concept of the CTO Journal, where the discipline of writing helps you clarify your voice and role. Using this journal is the perfect opportunity to look into what you have written and, if you haven't started yet, to start writing. If you do not know where to start, you can begin by asking the question: *What do I need to stop doing to become the CTO I am meant to be?*

The CTO Role Canvas is your tool for expressing that fixed role in a way that fits your leadership style, strengths, and current context. It helps you visualize not just what you are responsible for but also how you will engage with responsibility. This is not about redefining the CTO position; it is about executing it with clarity, intention, and sustainability.

Use it as a working document, something you revisit, revise, and refine over time. This canvas is not just about tasks. It's also about energy design, value alignment, and systems awareness. You are no longer a technology lead; you are designing the system that creates leadership flow, not leadership friction.

Here is how to begin:

- **Start with your current map.** What do you spend time on right now? The CTO Levels™ Framework is a great tool to learn what you *should* be working on at this level of complexity. Write it all down.
- **Mark what energizes and drains you.** Circle the high-impact, high-energy tasks. Cross out the ones that exhaust you.
- **Overlay your aspirations.** What work makes you feel most alive? Where do you create the most strategic leverage?
- **Identify where your current responsibilities clash with your intended role.** These are negotiation zones, not failure points.

At the same time, while it is important to consider that you will be shaping the role to have the best performance, some parts of the job will need to get done regardless of whether they are energizing or draining. In this case, it will help to delegate some of these tasks to other team members. We will address this issue further along in other chapters, where you will learn how to optimize your time and the tasks that do not require your specific attention.

Designing your role is about more than reclaiming time; it's about reclaiming purpose. A lot of the energy you spend gets drained by the unknown or uncertainty about what should be worked on. Once you have mapped where your energy flows and where it gets blocked, it is time to pinpoint where your contribution creates the most meaningful impact. This is where clarity turns into leverage by identifying your "value creation zones," the areas where your unique talents meet your company's deepest needs.

Value Creation Zones

One of the most powerful keys in role design is identifying your *value creation zones*, the intersection between what you love doing, what you are great at, and what the business desperately needs to deal with its complexities and operate within the expected levels. Too many CTOs get trapped doing what they are *good at* rather than what they were *meant for.* Often, the value they create isn't visible until it's missing. Your goal is not to chase value in every direction but to find the places where your contribution is exponential, not just functional.

Ask yourself these questions:

- *What outcomes improve significantly when I am involved?*
- *What feels easy to me but drives breakthroughs for others?*
- *What recurring problems or blind spots do I instinctively notice?*

Use this to identify your three to five personal value zones. These are the pillars of your authentic CTO expression. When you lead from here, your organization moves faster, with more clarity and less friction. You are no longer in reactive motion. You are driving strategic change.

Energy Allocation: The Invisible Blueprint

A CTO's real calendar is not measured in meetings but in *energy* (Martell, 2023). One of the greatest risks in high-performance leadership is energy misallocation: spending your best mental and emotional bandwidth on low-leverage work. Over time, this drains your calendar and your confidence.

But this calendar-as-energy metaphor gets sharper when you consider how fragile concentrated energy actually is. Field studies show that when knowledge workers are interrupted by a message, a question, or a quick Slack ping, it does not just pause the clock: It also affects deep work, when you need your complete focus. On average, it takes nearly 23 minutes before a person can fully refocus on their original task after an interruption (Mark et al., 2008). Yet, modern information workers often switch tasks every few minutes. In one of Mark's earlier studies, workers averaged about 3 minutes on a task before switching.

For a CTO, especially one whose highest-leverage contributions lie in strategic thinking, architecture design, long-term planning, or mentoring, this presents a structural risk: Constant context switching does not just fragment the schedule; it also fragments the mind. If every small interruption triggers a 20-plus-minute recovery cost, cumulative interruptions can leave you chronically unfocused, reacting frantically rather than building deliberately (Mark et al., 2008). The result is not just lost productivity but also fatigue, cognitive drain, and poor output quality.

Recognizing this changes how you treat your time and energy: not as interchangeable units, but as a limited reservoir needing protection. If you are serious about defining your CTO identity (allocating where you deliver value and where you delegate), you must design your own boundaries. Batch communications, block long focus sessions, create "no interruptions" windows. By doing so, you preserve the attention needed for high-impact thinking and avoid the silent tax that context switching imposes on every critical decision you must make.

Here is a simple audit you can run:

- Review last week's calendar.
- For each hour, for example, highlight in green the activities that were energizing, in orange those that were draining, and in purple those that were neutral.
- Add up the time spent in each category.

Now ask

- Are your best hours going to your highest-leverage work?
- Where are you compromising? Why?
- What must change for you to be in your zone of genius more than 50% of the time?

This exercise is not just about optimization. It is also about integrity. If you say your role is strategic but your energy goes to status updates and firefighting, something's out of sync. And that misalignment has a cost in culture, delivery, and well-being.

The Negotiation: Rewriting Your Responsibilities With the Business

Designing your ideal set of responsibilities is not a solo act. At some point, you will need to negotiate. This does not mean you will be fighting for your job, at least not in all cases. You will be cocreating a version of it that works for both *you* and *the business.* This can feel uncomfortable, especially if you have internalized the idea that being constantly available is the same as being valuable. But here is the truth: *Clarity brings confidence for everyone.*

When your executive peers understand what you are optimizing for, they are more likely to respect your time, protect your boundaries, and align outcomes. Here is a three-step approach to renegotiating your role:

1. **Frame the conversation strategically.** Lead with outcomes, not preferences.
2. **Share the evidence.** Use the CTO Levels™ Framework, your calendar audit, and value zones to show how you have reflected on what the business needs.
3. **Invite collaboration:** Ask: "How do you see my role evolving as we grow? Where can I lean in deeper, and where might I be overfunctioning?"

Negotiation is not confrontation; it should be seen more as an alignment ritual. When done with curiosity and confidence, it leads to better performance, not political risk.

The Liberation: Boundaries That Create Impact

In my work with CTOs, I noticed that several of them resist boundaries because they have been conditioned to see them as selfish, rigid, or even dangerous. However, these are boundaries designed to protect your time and your production, not those that will isolate you from the team or other executives. Simple actions, such as saying "no" or delegating to others, should work as protection to ensure your time is well managed, not a barrier that will prevent you and the team from working.

In fast-moving companies, we are taught to say yes, stay visible, and be available—always. But here is the leadership truth that nobody teaches you early on: Boundaries are not barriers. They are filters that will not block your value but *concentrate* it. They do not make you less involved; they make your involvement matter more.

A major reason CTOs struggle to protect their energy is that many are used to seeing boundaries as optional or even harmful. But the research on context switching paints a very different picture. According to different studies, the global economy loses billions per year to the cognitive drag of constant task shifting (Parikh, 2022).

With boundaries, your time becomes a curated environment where high-leverage thinking can occur. The irony is that the more senior you are, the more your organization depends on your depth, not your availability. Boundaries are what make this depth possible and protect your genius. They carve out space for the strategic thinking, deep focus, and vision alignment your team *actually* needs from you but often cannot access when you are buried in tactical noise. Without them, your time becomes a landfill of low-leverage obligations. With them, it becomes a garden of intentional impact.

One of the most liberating milestones I see in CTO coaching is when a leader realizes that they do not need to be the center of every conversation, especially once they have built a team capable of moving forward without their daily input. Letting go does not mean you are disengaged; it means the system you designed is finally doing its job.

It's a mindset shift

- from *presence* equals *relevance* to *absence* equals *trust.*
- from *proving value* to *designing leverage.*
- from *controlling outcomes* to *shaping environments.*

Sometimes, your absence is the most powerful signal of your leadership maturity.

One CTO I worked with recently had an uncomfortable moment when his Product team delivered a high-stakes road map presentation to the executive teams without including him. His instinct was frustration: *Why didn't they loop me in?* But after some reflection, he realized what had actually happened. His team had internalized the strategy. They were thinking holistically. They were owning the outcome. They did not need him, and that was not a failure but a success by design.

In this case, you can work within certain scopes:

- Are you going to call people out?
- Are you going to ask for an update?
- Are you going to pout that you were left out?

The fact is that the team shared information, and if you need that information, ask for it without frustration about not being kept in the loop. I believe this is a key breakthrough in leadership: when we can develop the ability to not feel left out and, with intention, include ourselves where necessary.

Liberation is not about doing less for the sake of it. It is about doing less of what distracts you so you can do more of what matters. When you protect your focus and design boundaries, you free your team to rise and your role to scale.

Refactor in Action

The leadership team was knee-deep in quarterly planning when the CEO leaned across the table. "Mira, you're in charge of way too much. Half of these projects don't even need a CTO."

Mira felt her face flush. For months, she had been everywhere—in sprint reviews, budget debates, and late-night email threads. She had convinced herself that being indispensable meant she was doing well. But suddenly, it sounded like failure.

On the drive home, the words kept echoing. *Too much. Doesn't need a CTO.* She wanted to argue, but for the first time, she asked herself, *What should a CTO actually be doing here?*

That night, instead of collapsing into bed, Mira opened her calendar and started marking each meeting: green if it created real leverage and red if someone else could handle it. By the time she finished, the page glowed red.

The next morning, she did something uncharacteristic: She pulled her CEO aside. "If you want me to build this company's future, I can't spend my days chasing tickets and vendor quotes."

He raised an eyebrow, skeptical. "So, what does that mean?"

She didn't have a neat answer yet—but she had proof. She showed him the calendar audit, the flood of red blocks, the thin green slices where she actually operated as a CTO. It wasn't perfect, but it was honest.

The conversation was awkward. It ended with more questions than answers. But it cracked something open. For the first time, Mira wasn't defaulting to what everyone else needed. She was negotiating what the role should be.

It wasn't neat. It wasn't resolved. But it was the first time Mira stopped inheriting her role and started designing it.

REFLECTION PROMPT

What is one small shift you could make this week to move closer to your zone of genius? And what would become possible if you did that consistently for the next 90 days?

How to Start Redesigning Your Role

You do not need permission to begin. You just need a window of clarity and the willingness to act on it. Here are four steps to help you start today:

1. **Audit your energy:** Review the past two weeks of your calendar. What energized you, what drained you, and what didn't need you?
2. **Draft your CTO Role Canvas:** Don't overthink it. Start with what is real. Then, map what is ideal.
3. **Choose one boundary to establish:** Maybe it's a standing meeting you no longer need to attend. Maybe it's a block of deep work time you fiercely protect.
4. **Have one strategic conversation:** Whether with your CEO, cofounder, or team, make your thinking visible. Show your intent to be a more focused, effective CTO, not a reactive and stretched-thin one.

This is not about perfection. It is about how you engage with a role that is already yours *and* already defined. The CTO's role has not changed. But how do *you* bring it to life? That's where your power is. The most effective CTOs are not the ones who try to do everything. They are the ones who know exactly where they create the most value and dare to design around it. In the next chapter, we will move beyond role design and focus on *alignment*: how to connect your work to the broader business, build strategic trust with your executive peers, and lead with clarity across functions.

CHAPTER 3

The Alignment Engine: Becoming a Strategic Partner

You can't scale leadership without alignment.
And you can't create alignment without clarity.

You are leading the engineering organization. You have a vision, a hiring plan, a road map, and maybe even a decent product pipeline. But something still feels off:

- Your CEO is anxious about timelines.
- Your Product peer is frustrated about velocity.
- You are getting pulled into strategy conversations late (or not at all).

You wonder, *How can I be doing everything right but still feel like I am not fully trusted?* The hard truth about being technically competent and reliable is that it is not enough. Not at the executive level. To fully inhabit your CTO role, not just in title but also in influence, you need to build an alignment system: a deliberate structure for creating trust, shared context, and forward momentum between you and the rest of the business.

That system starts with *you,* and it scales through clarity.

Alignment Is a System, Not a Vibe

Many CTOs I talk to confuse alignment with agreement. They chase harmony, thinking that if everyone nods in the meeting, the company is aligned. But alignment is not about harmony or a feeling. It is a *system*. Additionally, having everyone agree to what you are saying is not alignment either; this is consensus seeking. And it often leads to frustration, resentment, or miscommunication when priorities shift.

A system implies *inputs, processes,* and *outputs.* It's not passive, it's engineered, and, as the CTO, you are not just part of that system; you also help *design it.* True alignment is not about getting a buy-in for your ideas but about creating a shared understanding, and, like any system, it needs structure. In our coaching at Coach EXE and 7CTOs, we often say, "Alignment isn't one or two great meetings, it's a series of hard meetings. It's a rhythm."

Finally, you should remember that alignment is not *what* you communicate but *how often, with whom,* and *through what lens.* Alignment is the engine that powers everything else: trust, clarity, autonomy, and momentum. Without it, you are managing tension. With it, you are cocreating a direction.

Take a moment to reflect on the tensions you have felt within your Engineering teams: missed expectations, repeated debates, slow decision-making, or quiet disengagement. Ask yourself whether those moments were truly about capability or effort, or whether they stemmed from misalignment: unclear priorities, mismatched incentives, or different interpretations of success. When you trace tension back to its source, you will often find that alignment was never fully established. Recognizing this is the first step toward replacing friction with shared direction.

Start With Strategic Self-Awareness

Before you start aligning with others, you have to be honest with yourself. The first misalignment happens *internally*, between what you think your role is and how others experience you in it. Ask yourself the following questions, which you started scratching the surface of in the previous chapter:

- *Am I showing up primarily as a builder, operator, or strategist?*
- *Does my calendar reflect what the company actually needs from me?*
- *Have I made my vision, values, and boundaries visible? Or am I assuming others just "get it"?*

This internal clarity is the first gear in the alignment engine. If you are unclear, everyone else will be guessing, or maybe worse, projecting their own assumptions onto your role. Remember that you are not here to *chase* alignment. You are here to *generate* it through clarity of purpose and consistency of presence.

The CEO–CTO Relationship

Your relationship with the CEO is the single most important channel in your world. When it's strong, it becomes the propeller of confidence and clarity for everyone else. When it's weak, even the best project strategies and the strongest Engineering teams can quietly suffer. Misalignment here trickles down fast: priorities get scrambled, decisions slow, and trust erodes across the organization.

The value of the CTO–CEO alignment is not in reporting how Engineering is performing in isolation but in explaining why those signals matter to the business. This is where something like the DORA metrics demonstrates what a powerful connection it could be. DORA's research is based on data from over 31,000 professionals across multiple years and identifies 4 metrics that consistently correlate with high-performing engineering organizations (Harvey, 2026). However, their real impact emerges only when the CTO connects them directly to business agility and results, acting as a translation layer between technology and business outcomes.

DORA metrics act as proxies for how quickly and safely a company can turn strategy into execution:

- **Deployment frequency:** Faster delivery of customer value, enabling quicker revenue realization and more rapid experimentation
- **Lead time for changes:** Greater responsiveness to market shifts and customer feedback, reducing the costs of delay
- **Change failure rate:** More predictable execution, fewer customer-impacting defects, and less revenue loss to rework or outages
- **Mean time to recovery (MTTR):** Lower operational risk, reduced downtime costs, and increased customer trust during inevitable failures

Simply telling a CEO that the deployment frequency has increased or MTTR has improved is insufficient if those metrics are not grounded in business impact. It is the CTO's role to make that connection explicit and to show how Engineering performance translates into business growth, resilience, and a competitive advantage. When these metrics are aligned with business outcomes, they become shared indicators of progress rather than siloed technical measurements.

As you can see, there is a clear and direct link between engineering execution and business outcomes, but that link only stays strong if the broader leadership is aligned. A CTO whose vision, priorities, and capacity are synchronized with the CEO's helps ensure that technical strategy supports business goals, and that resources (time, budget, and people) flow where they actually matter. When that alignment falters, expectations diverge: The Engineering team may optimize for velocity, but the business may need stability; or the business pushes for features, but the architecture demands caution. The result is degraded DORA metrics, slower delivery, more failures, and, ultimately, weaker business performance, even if the engineers are working hard.

In essence, a strong CEO–CTO partnership is not just nice to have; it is also a strategic lever. By investing in clarity and trust at the top, you enable your teams to hit the DORA targets that research shows double the odds of achieving business success (Harvey, 2026). In doing so, you protect your organization from the silent cost of misalignment: squandered time, wasted cycles, and a lost competitive advantage.

A CEO does not need you to echo their words or act as a rubber stamp. They also do not want a purely tactical executor. What they need is a multiplier, someone who can take the company's objectives and transform technology products or services through resilient and adaptable technology. That multiplier role only emerges when there is space for both trust and tension. Trust allows the CEO to step back, knowing you are steering your ship. Tension ensures the CEO's assumptions are being challenged while you advocate for long-term stability and bring a tech perspective into decisions that affect the whole business.

But this partnership can't be left to chance. Strong CEO–CTO relationships are designed, not assumed. Start by asking yourself

- *How often do we meet, and are those meetings actually strategic, or are they just status updates?*
- *Do we share the same definition of success for Engineering, beyond hitting deadlines?*
- *Do I truly know what keeps my CEO awake at night and how technology can address or prevent those issues?*

A CTO operating purely in the Engineering section is replaceable. A CTO who operates as a trusted business partner, one who can translate objectives into scalable systems, is indispensable. Alignment at the executive level is absolutely not "staying in your lane." It's about seeing the whole landscape and ensuring that the machine you are building is tuned to the company's real trajectory, not just your team's road map.

Make the Invisible Visible

Executive friction does not come from ego or sabotage, although this is certainly possible. It comes from opacity. As the CTO, you live with complexity others do not see: subtle architecture dependencies, technical debt that silently compounds, and a culture that could snap under another rushed deadline. When these dynamics remain invisible, your peers will unknowingly make decisions that undermine the system. This happens not because they are careless but because they do not know what is at stake.

Google's Site Reliability Engineering (SRE) research reinforces just how costly this invisibility can be. After analyzing thousands of postmortems over seven years, Google found that most incidents were not caused by isolated code defects but by systems-level factors: configuration errors, flawed rollout processes, hidden dependency chains, and unclear ownership boundaries (Warner et al., n.d.). In other words, the failures that make headlines rarely start as engineering mistakes; they start as organizational blind spots. When leaders outside Engineering do not notice those systemic pressures, they will unintentionally green-light decisions that amplify fragility.

Salesforce Engineering and other large-scale platforms have reported similar patterns. Their incident analyses show that outages almost always have multiple contributing factors, and that human error is rarely the root cause on its own. Instead, it is the ecosystem around the work (rushed timelines, unclear responsibilities, brittle integrations, and competing priorities) that sets the stage. These dynamics mirror what CTOs face daily: complex systems in which technical choices and organizational behaviors are inseparable. If those interdependencies stay unspoken, they become landmines scattered across the business.

This is why surfacing complexity is an act of leadership, not defensiveness. Your peers do not need every flow diagram or incident log, but they do need clarity about systemic risk: what could break, why it could break, and what trade-offs they are implicitly making when they push for speed, scope, or shortcuts. When you illuminate those patterns not as excuses but as operational truths, you give the executive team the situational awareness needed to make decisions that strengthen rather than strain the system. Only then can they fully understand what is at stake and operate as true partners rather than accidental saboteurs.

Your job is not to bury them in jargon or shield them from every technical detail. It is to translate complexity into consequence so they can make informed, confident decisions. This does not mean dumbing things down but rather contextualizing them. Here are some examples to consider:

- Instead of saying, "We need to refactor the API layer," explain, "If we don't, every future feature will take 50%–75% longer, and our developers will burn out from frustration. This is about protecting speed and morale, not just code quality."
- Instead of saying, "The architecture needs to evolve," try saying, "We can handle 10 enterprise clients with our current system. But our sales plan calls for 50 by year end. Without evolving the architecture, we will stall growth. This will happen not because we lack customers but because we cannot serve them."

This is systems thinking in action. You are not just fixing code; you are also narrating the ripple effects of the system so others can see why your recommendations matter. This ability to make the invisible visible is one of the fastest ways to earn lasting strategic trust.

Tension Is Not a Sign of Failure

Many CTOs I talk to feel drained by conflict, as if every disagreement signals dysfunction. In reality, every thriving executive team lives with a baseline of tension. If there is no tension, chances are people are avoiding the hard conversations. For a CTO, that tension is often amplified because you sit at a crossroads of competing needs:

- Product is pushing for speed to hit market windows.
- Sales is pressing for promises to close deals.

- Finance is asking you to do more with less.
- Meanwhile, you are responsible for ensuring stability, scalability, and a sane culture.

This does not mean you are failing. It simply means that the system is alive. Friction becomes dangerous only when it goes unmanaged or unnoticed.

Your role is not to erase the tension; it is to hold the space where trade-offs are surfaced and decisions are made consciously. Advocate for long-term resilience, but stay open to calculated risks. Clarify the real costs (technical, cultural, and financial) so that no one is making blind choices.

In high-functioning teams, tension becomes creative fuel. It sparks smarter decisions and more sustainable strategies. In low-trust environments, it turns into political drag, where fear and finger-pointing dominate. The difference lies in how you, the CTO, show up. If you approach friction as a part of the system, you transform tension into forward motion.

This perspective echoes Peter Senge's (2006) work in *The Fifth Discipline,* where tension is treated not as a problem to eliminate but as feedback to understand. When leaders view tension as information, such as signals about misaligned goals, broken feedback loops, or unspoken constraints, it becomes a diagnostic tool rather than a threat. Instead of suppressing disagreement, high-performing organizations learn from it.

Seen this way, tension is not a leadership failure; it is evidence that the system is trying to tell you something. The CTO's role is to listen, interpret, and respond thoughtfully, and turn this feedback into learning and learning into alignment. When tension is handled this way, it stops slowing the organization down and starts pulling it forward.

From Tech Advocate to Systems Leader

There is a moment in every CTO's evolution when they stop defining themselves by technology alone and start embracing their role as a steward of the whole business system. This is the point where true alignment clicks. Before this shift, your energy is spent reacting: protecting your team, defending technical decisions, and constantly justifying your worth.

After this shift, you are helping shape priorities at the table, trusted not just as "the tech person" but as someone who sees the interplay between revenue, risk, culture, and capability. This is not about ego or title. It is about how you show up and react:

- You stop reacting to decisions and start shaping them early.
- You stop repeating your value and start delivering it in ways the business feels, not just sees.
- You lead conversations about growth, trade-offs, and opportunity with the same confidence as you lead architectural decisions.

And it all points back to the CTO's fixed core:

- Understand the company's objectives.
- Understand and steward the technology budget.
- Build the organization that builds the technology.

Master those, and you are not just leading Engineering; you are also leading *with* the business.

Refactor in Action

The product road map review was dragging into its third hour when Eric noticed the CEO's arms crossed, his face tight with frustration.

"Why am I just hearing about this now?" the CEO snapped, pointing to the slide about a delayed integration. "This changes our whole sales plan."

Eric froze. The engineering work was on track. The code was solid. The delay had nothing to do with execution—it was a trade-off he thought was obvious. He had pushed the team toward stability instead of speed, assuming the value spoke for itself. Apparently, it hadn't.

After the meeting, his Product counterpart pulled him aside. "Look, Eric, we need you in these conversations earlier. I can't keep translating what you're optimizing for."

It stung because she was right. Eric was operating in what he called "quiet execution mode." Head down, delivering, confident that results would earn trust. But results without context weren't enough. To his peers, it looked like he was hiding the ball.

That week, he tried something different. Instead of waiting for the next blow-up, he drafted a short "alignment brief" for the CEO and product lead. One page, no jargon: what was on track, what trade-offs had been made, and what risks were looming if priorities didn't shift.

He hit send and braced himself.

The response surprised him. The CEO replied within minutes: "This is exactly what I needed. Let's talk through the risks tomorrow." Product followed up too: "This helps so much—thank you for putting it in business terms."

It wasn't magic. There were still tensions, still debates. But now, those conversations were happening with shared context instead of crossed arms and blindsides.

Eric realized alignment wasn't about being agreeable or "delivering silently." It was a rhythm—making the invisible visible, consistently, so that trust could compound instead of erode.

Create an Alignment Rhythm

Alignment isn't a one-off fix. It's a living rhythm that keeps the system coherent as it evolves. Without rhythm, misalignment creeps back the moment priorities shift or the quarter heats up. Here is a strategy that works for many CTOs:

- **Weekly:** CEO 1:1s, focused on outcomes and risks, not just status.
- **Biweekly:** Product, Engineering, and Design alignment check-ins, ensuring shared priorities and fast blocker removal.
- **Monthly:** Engineering communicates business impact, such as what new capabilities were unlocked, and not just velocity metrics.
- **Quarterly:** Cross-executive strategy sessions to reassess objectives, budgets, and key trade-offs for the next cycle.

The way you structure your agenda will vary, but the principle is universal: *Make alignment predictable, visible, and focused*. When communication is rhythmic, your workdays flow.

Realignment as Leadership Maturity

Even with a well-built alignment engine, drift happens. Companies evolve. Priorities shift. A new executive joins the team. Growth strains systems that once felt solid. When that happens, many CTOs see misalignment as failure. They feel like they have dropped the ball. However, you must think of misalignment not as such, but as a signal. The most mature leaders don't avoid misalignment; they detect it early and realign quickly.

Realignment begins with awareness: noticing subtle cues that trust or clarity is slipping. Maybe the CEO starts double-checking your decisions. Maybe Product is bypassing Engineering to get faster answers. Maybe your calendar is once again dominated by tasks that don't belong in your zone of genius. These are not annoyances to brush off but indicators that the system needs tuning.

Then comes action. Mature CTOs do not wait for frustration to boil over. They

- revisit what success looks like for their role and confirm it with the CEO.
- request an updated business context so they can prioritize correctly.
- bring to the surface ambiguities or tensions directly, before they metastasize.
- reaffirm the link between technology, revenue, and growth so peers see the full picture.

And perhaps most importantly, *they tell the truth*. They speak up when trade-offs are being unconsciously made, when expectations have shifted without discussion, or when their role needs to flex to keep the company healthy. You cannot lead a system you are afraid to confront. And you cannot be a strategic partner if you are silently compensating for the dysfunction. Realignment is not a retreat; it's an assertion: "I am here to build the machine that builds the future. Let's make sure we are pointing it in the right direction, together."

The CTO Alignment Kit

Use this checklist as a regular self-audit. If you cannot check most of these boxes, it's time to rebuild your alignment engine:

- ☐ I can clearly articulate how Engineering supports the company's objectives.
- ☐ My CEO understands how our tech decisions impact revenue, risk, and growth.
- ☐ My Product counterpart and I share clarity on trade-offs and decision-making authority.
- ☐ My calendar includes strategic thinking time, not just operational firefighting.
- ☐ I have a predictable rhythm for surfacing complexity in a business-friendly way.
- ☐ I am not dodging tension; I am using it to help the team make informed, intentional decisions.

If you marked fewer than four, that is not failure; it's feedback. Your alignment engine needs a tune-up. That might mean reestablishing CEO rhythms, improving how you translate complexity, or reinforcing decision cadence across your peers. Remember: *Alignment is not a one-time project. It's an ongoing discipline.* Treat it like system maintenance. Ignore it, and the friction builds up until the machine slows or breaks. Nurture it, and the whole company runs faster and smoother.

REFLECTION PROMPT

Where is alignment weakest in your world right now? What single conversation could begin to fix it?

Alignment Is Not Optional; It's Your Leverage

You can't scale leadership without alignment. You can't build trust without clarity. You cannot operate as a true peer on the executive team if you are trapped in reaction mode, constantly translating or justifying your work. The CTO's core role is fixed, but how you live that role and how you align others around it will determine whether you are a bottleneck or a force multiplier. You are not just here to lead technology. You are also here to lead *with* the business, and alignment is the engine that allows your leadership to scale beyond the boundaries of Engineering.

Clarity and alignment free you from firefighting and position you as a true strategic leader. But alignment alone isn't enough if you are still working at the mercy of your calendar—scattered, drained, and reactive. In the next chapter, we will tackle the most overlooked system a CTO needs to master: time. Not as a scheduling tool, but as a leadership system you can design to protect your energy, amplify your leverage, and keep you operating at your highest value.

CHAPTER 4

The Firefighter's Escape: Breaking the Reactivity Cycle

If your value is measured by how many fires you put out, you'll never build the firebreaks that prevent them.

I'm sure you didn't imagine spending your days running from one crisis to the next. When we first step into this role, we picture ourselves as architects, shaping scalable systems, inspiring teams, and aligning technology to the company's vision. But somewhere along the path, often without even realizing it, we morph into something else entirely.

Suddenly, the calendar is a minefield of mindless monthly and weekly meetings. A production issue before your morning beverage. A customer escalation midafternoon. An annoying text as you are sitting down for dinner. You spend your days not building, not coaching, not designing the future, but constantly *reacting*; running from flare-up to flare-up, exhausted. And yet, you feel oddly validated.

Therein lies the trap. There is a part of the crisis mode that feels good. It offers a hint of clarity and purpose. In the chaos, no one is questioning your value. You are the hero, the one who swoops in and saves the company. It feels decisive, necessary, even noble. However, the truth is that this cycle is slowly eroding everything that truly makes you valuable.

There is another, more subtle dynamic at play as well. Crisis-driven leadership can unintentionally appeal to the CEO's sympathetic instincts, turning chaos into a form of cover. When progress stalls, attention shifts away from outcomes and toward explanations: team conflict, unexpected dependencies, system upgrades, and personal strain. These factors are often real and deserve empathy.

However, over time, this pattern can become a way of deflecting accountability, where shared understanding is replaced by shared concern. The CTO is no longer measured by commitments kept but by how convincingly the struggle is conveyed. What begins as transparency slowly turns into a habit that erodes trust, alignment, and the credibility required for long-term leadership.

Every time you put yourself at the center of an emergency, you teach your team to wait for you instead of solving problems on their own. Every time you race to fix something, you rob your organization of the chance to build systems and processes that would prevent the problem from recurring. You become indispensable in the short term, but at the cost of becoming a bottleneck and a liability in the long term.

This is what I call the *firefighter's trap*. The more chaos you handle, the more chaos your company allows (or even creates) because subconsciously, everyone knows you will show up with the hose. Your adrenaline tells you you're leading. In reality, you are just perpetuating a system that cannot run without you.

Escaping this trap is not about ignoring the urgent or turning a blind eye to real threats. It's about changing your relationship to urgency itself: understanding what fires truly matter, which ones can be handed to others, and which ones should not exist at all. It's about trading heroics for architecture and shifting your energy from the thrill of putting out flames to the slower, quieter, but far more powerful work of building a company that does not ignite so easily.

The Adrenaline Addiction

One of the hardest parts about breaking free from constant reactivity is acknowledging how much we come to crave it. Few people will admit this out loud, but most technology leaders I coach will quietly nod when I ask, "Do you secretly enjoy the chaos?" The main reason is that in the whirlwind of crisis, there is certainty. A major outage does not need a strategy deck or a lengthy discussion. The course of action is obvious. Decisions are immediate, and when you deliver the fix, everyone notices. You are visible. Praised. Gratified.

Psychology helps explain why this cycle feels so seductive. High-pressure environments trigger a biochemical cocktail of adrenaline, cortisol, and dopamine that sharpens focus and delivers short bursts of heightened performance. Studies on crisis addiction in corporate leaders note that urgent, high-stakes tasks activate the brain's reward circuitry far more intensely than long-term strategic work (Ripla, 2025).

This creates a reinforcement loop: The more crises you handle, the more your brain seeks the neurochemical payoff that comes with immediacy, visibility, and praise. Organizational-behavior research on "hero culture" echoes this pattern, showing how leaders become unintentionally conditioned to chase fires because firefighting is publicly rewarded while slow, preventive, systemic work is often invisible.

This also aligns with decades of productivity research, including well-known frameworks like the Eisenhower Matrix, which highlight how the human brain instinctively prioritizes urgent-but-less-important tasks over important-but-not-urgent ones. In crises, the choices feel simple, and the feedback is instantaneous. Strategic work, in contrast, offers no immediate dopamine hit: Its rewards are delayed, diffuse, and rarely celebrated. Neuroscience findings further suggest that chronic cortisol spikes from operating in constant crisis mode may enhance short-term execution but steadily erode long-term cognitive capacity, decision quality, and emotional regulation (Priya, 2023).

In other words, the very habits that make a leader feel effective in the moment can undermine their effectiveness over time. This creates a psychological trap for many CTOs: The more they excel in chaos, the more their brains (and often their organizations) push them back into it. Without awareness, reactivity becomes identity.

For many CTOs, that visibility scratches an itch that the quieter parts of leadership, such as the planning sessions, the mentoring conversations, and the months-long architectural initiatives, often don't. Those things don't produce a dopamine rush. They don't lead to frantic thank-you messages or CEO shoutouts. They are slow, deliberate, and, at times, thankless, potentially leading to conflict avoidance, whereas, in a crisis, hard work is needed.

And so, without realizing it, you start to gravitate toward the chaos. Each fire reinforces your sense of purpose. Each save fuels your reputation as a hero. Meanwhile, the real work, the work that would actually make the organization stronger, more resilient, and less dependent on you, gets postponed.

This is how smart, capable leaders burn out. It is also how promising companies plateau, because their technology function, while responsible, never grows into a truly scalable machine. The first step toward breaking the cycle is acknowledging that the adrenaline high is a trap; that it is not a measure of your value, but a warning sign that you have become too central to too many things.

Seeing Fires for What They Really Are

But how do we know if we are caught in the adrenaline loop? Every ping feels urgent. Every interruption seems like a crisis waiting to explode. However, if you want to regain control of your time, energy, and role, you must learn to see these fires for what they truly are.

Some fires are real *infernos*: an outage, a breach, a failure that immediately threatens customers, revenue, or the company's reputation. Those require your direct involvement, not because you are the only one who can act, but because your presence provides clarity and calm when the stakes are existential.

Other fires are more like fireplace fires: moments when executive priorities clash, road maps get pulled in conflicting directions, or sales commitments outpace delivery capabilities. Those don't need you rushing in with a fix, but they do need your steady hand to realign the organization so that the same conflicts don't repeat endlessly.

In other types of fires, such as campfires, there are operational snags such as bugs, bottlenecks, or team conflicts that feel urgent but are well within your leaders' ability to resolve, provided you have given them the clarity and authority to do so.

And finally, there are the phantom flames, which include the anxieties, overreactions, the "this is on fire" Slack messages that turn out to be misunderstandings, incomplete data, or problems already being solved elsewhere. These moments rarely require executive intervention. What they need is context, perspective, and occasionally a calm signal from leadership that not all flickers of flames warrant escalation.

This is precisely why I created the CTO Compass (www.ctocompass.com). It was modeled as a practical framework to help CTOs stay slow, steady, and intentional while still moving the business forward. Instead of reacting to noise, the Compass provides a consistent way to evaluate interruptions against business objectives, organizational health, and leadership responsibility. By viewing each situation through this lens, you can deliberately choose whether to engage, delegate, coach, or step aside without losing momentum or trust.

Over time, using the CTO Compass does more than reduce reactivity. It exposes the underlying patterns that create chaos in the first place: fragile systems, underempowered leaders, unclear ownership, or misaligned expectations with peers. By responding consistently and calmly, you help extinguish false alarms and strengthen the organization's ability to operate with clarity, resilience, and purpose.

Building the Firebreaks

Once you understand where the fires are truly coming from, the real work begins: building the firebreaks that prevent sparks from turning into infernos. Most recurring crises don't originate in the code. They originate in the *system* that includes how the company plans, communicates, and allocates resources.

In my experience, most preventable fires fall into four root categories of the CTO Sentinel—*Speed, Stretch, Shield*, and *Sales*:

- **Speed-driven fires** happen when the company moves so fast (or promises so much) that even hiccups become major events. A single late task cascades into chaos because the deadlines were already unrealistic. Preventing these fires isn't about slowing everything down. It's about creating planning rituals and buffers that keep urgency from metastasizing into panic.

- **Stretch-driven fires** occur when teams are consistently operating at or beyond their sustainable limits. When everyone is maxed out, mistakes multiply. Quality dips. Tempers fray. Suddenly, every small problem feels catastrophic because no one has the capacity to absorb it. These fires require structural adjustments such as hiring rescoping, or simply saying "no" to work that the system cannot handle without breaking.
- **Shield-related fires** are the interruptions and disruptions that creep in when Engineering is exposed to every stray demand from other departments. It includes fighting a fragile infrastructure, volatile developer relationships, unclear direction, and an information or data overload without the ability to analyze it. It is more about external and internal threats that have to be staved off.
- Finally, **sales-driven** fires come from commitments made without clarity or coordination. It is that moment in which you hear, "We already told the customer that the feature will be live next week," that moment that throws the entire system into scramble mode. These can only be prevented by tightening the feedback loop between your technical teams and the people promising deliverables, and by standing firm when promises exceed your system's capacity.

Each of these categories has its own levers to pull. Some require better planning cadences. Some require deeper leadership bench strength. Some require renegotiating norms with your peers and the CEO. But all of them share one truth: *Fires do not stop because you move faster with a hose. They stop when you change the system that keeps sparking them.*

The Discipline of Strategic Neglect

There is a skill every CTO needs to master if they are going to escape the firefighter cycle: the ability to *not* respond. This feels counterintuitive, even reckless, for leaders who pride themselves on reliability. But strategic neglect isn't laziness. It's discipline. It's knowing that every time you jump into a noncritical issue, you are not just burning your own time; you are also stealing a growth opportunity from your leaders and reinforcing the dependency that keeps your calendar full of chaos.

Letting a director handle a tricky escalation, even if they stumble, teaches them far more than watching you fix it. Ignoring a phantom flame, the "urgent" issue that turns out to be a misunderstanding, trains your peers to pause and assess before summoning you. Allowing certain problems to breathe, rather than immediately fixing the symptom, often reveals the root cause that actually needs solving.

A growing body of organizational psychology research reinforces this idea that not responding can be a powerful developmental tool. Studies on managerial delegation consistently show that leaders who deliberately step back create conditions for autonomy, competence, and ownership to grow (Mathebula & Barnard, 2020). These are three ingredients considered central to self-determination theory. When senior leaders intervene too quickly, teams experience what researchers call learned dependence, where capability stagnates because the path of least resistance is waiting for the expert to jump in (Zimmerman, 1990). Strategic neglect, when applied thoughtfully, becomes a cycle-breaking mechanism.

Cognitive science also supports this practice. Studies on attention switching and cognitive load demonstrate that constant context-shifting (especially in response to low-impact interruptions) degrades problem-solving ability over time (Sherbino & Norman, 2021). Leaders who avoid reflexive intervention protect not only their focus but the organization's collective decision quality. Allowing noncritical issues to settle before reacting reduces noise in the system and increases the signal-to-noise ratio of true problems. In many environments, simply waiting 15–30 minutes before engaging materially reduces unnecessary escalations.

I have also observed evidence in operations and reliability engineering that supports calibrated nonintervention. Postmortem analyses with SRE and high-reliability organizations repeatedly show that premature fixes or leader-driven quick-saves often obscure the underlying contributing factors, delaying real resolutions. Allowing the team closest to the issue to lead the response produces more accurate diagnoses and more durable fixes. Strategic neglect is not passive but rather a form of leadership constraint designed to surface systemic truths that only emerge when you resist the urge to immediately solve.

Strategic neglect is about trust—trusting your team to rise, trusting the system you are building to hold on, and trusting yourself to stay focused on the few things that only you can truly do.

From Rescuer to Architect

The shift away from firefighting is not measured by how few emergencies exist. Fires will always happen. It is measured by how differently your world looks when they do. When you are still stuck in the firefighter loop, every spark pulls you in. Your calendar is shredded. Your evenings disappear. Strategic projects stall. You end the day exhausted, wondering if you have accomplished anything beyond survival.

Behavioral decision research affirms that there is a strong systemic tendency toward what psychologists call escalation of commitment: the inclination to persist, often unconsciously, in a course of action even when feedback shows it is failing (Keil et al., 2007). In those studies, decision-makers exposed to negative feedback rarely shift course; instead, they downplay problems, reinterpret the information, or overrationalize their prior investment. In practical terms, for a CTO, this cognitive bias may manifest as reflexively jumping in to "save" a developer issue, upgrade a feature, or patch a deployment, each time reinforcing a cycle where nothing improves structurally and every crisis becomes a self-fulfilling prophecy.

Research Evidence: The Escalation Trap

In the study "Why Software Projects Escalate: An Empirical Analysis and Test of Four Theoretical Models" (Keil et al., 2000), the authors surveyed information systems (IS) audit and control professionals and found that 30%–40% of IS projects exhibit escalation behavior, meaning that they commit to failing courses of action despite clear signs of trouble. These escalated projects showed significantly worse outcomes, with budget/schedule overruns, lower perceived success, and lower implementation performance, compared to nonescalation ones.

The study tested multiple theoretical explanations (self-justification, sunk-cost bias, prospect theory, and agency theory) and found them predictive of escalation behavior, meaning that the phenomenon is both widespread and systematic, not rare or anecdotal (Keil et al., 2000). But this is important for CTOs because it is not only about failed projects; it is about a behavioral bias that repeatedly leads decision-makers to pour more resources into failing work, often because of prior investment, pride, or lack of clarity. For a CTO prone to "rescue mode," this research demonstrates why rescue-driven escalation can embed fragility rather than resolve it.

Research Evidence: Breaking the Cycle

The paper "Escalation of Commitment in MIS Projects" aggregates decades of research across many studies, finding that escalation is a major explanation for why IT and software projects often exceed time and budget constraints (Desai & Chulkov, 2011). Rather than mere accidental mismanagement, escalation reflects a predictable pattern arising from structural, psychological, and organizational pressures, not from occasional bad luck.

These are not edge cases: Escalation and clinging to failing projects, perpetually firefighting, and refusing to pull the plug are deeply embedded in many software organizations. Recognizing this as a systemic bias rather than an occasional misjudgment helps frame strategic neglect (or refusal to escalate) not as passivity but as disciplinary disruption. Firefighting alone does not solve root causes: It only patches symptoms.

When you break that cycle and intentionally withhold immediate action to see whether the issue stabilizes, whether the team steps up, or whether underlying patterns emerge, you are applying the antidote to escalation bias. Letting some issues breathe, seeing which ones truly matter, and trusting your people or system to respond refocuses your bandwidth on what really counts. The result is that most issues never reach you.

The ones that do are genuine, high-stakes problems, not emotional escalations. Your leaders solve more on their own. Your CEO trusts you because they are informed, not because you are perpetually available. You regain time, not just for deep, strategic work but also for your life outside of work. This, in fact, no longer feels like something you can only attend at midnight or on rare, disconnected weekends.

The transformation doesn't just change how your company runs. It changes how you feel as a leader. Instead of being drained by the job, you are energized again. Instead of surviving each week, you are shaping the months and quarters ahead, and, in doing so, you are finally escaping the loop as you will be only responding to what truly matters.

Refactor in Action

The text came at 9:14 p.m.

"Do you have 5 minutes tomorrow? I need to talk about David."

Simone knew what was coming. Rachel, her VP of Engineering, had been clashing with David, her Director of Platform, for months. The friction wasn't about strategy or technical direction—it was personal. Different communication styles, old resentments, a misread comment in a planning meeting that had calcified into cold silence.

By the time Simone finished her coffee the next morning, she'd already spent 40 minutes listening to Rachel vent. By lunch, David had cornered her with his version. By end of day, she'd scheduled a three-way conversation to "clear the air"—the fourth such conversation in as many months.

It never stuck. The truces lasted a week, maybe two, before another flare-up landed in her inbox. Each time, Simone absorbed the tension, smoothed the edges, and sent them back to work. She told herself it was leadership. It felt more like couples therapy.

The pattern spread. When two senior engineers had a turf dispute, it came to Simone. When a director felt undermined by a peer, Simone got the call. She had become the emotional first responder for her entire leadership team—the person who held the anxiety so others didn't have to sit in discomfort.

Her own work suffered. Strategic planning got pushed. Her CEO started asking why the Engineering road map was always "almost ready." Simone's wife noticed she'd stopped talking about ideas and only talked about people problems.

The breaking point came during a quarterly offsite. Simone had spent two hours the night before coaching Rachel through yet another David incident. The next morning, she watched them sit on opposite sides of the room, arms crossed, waiting for her to fix it. And she realized: They weren't learning anything. They were just waiting for her to absorb the tension again.

That afternoon, Simone tried something different. When Rachel pinged her about David's "undermining behavior" in a planning discussion, Simone didn't offer to mediate. Instead, she replied: "That sounds frustrating. What's your plan for addressing it with him directly?"

Silence. Then: "I was hoping you could talk to him."

"I think this one needs to be between you two. I trust you to handle it."

The next few weeks were uncomfortable. Rachel was annoyed. David seemed confused when no one intervened. A conversation happened between them—awkward, incomplete, but real. It wasn't perfect. But it was theirs.

Simone created a simple boundary: She would coach her leaders on how to navigate conflict, but she would no longer carry the conflict for them. When someone came to her with peer tension, she asked three questions: Have you spoken to them directly? What do you need from that conversation? How can I support you in having it?

Some leaders rose to it quickly. Others struggled and needed more coaching on the skill itself. But the dynamic shifted. Tension still surfaced—it always does—but it no longer funneled automatically to Simone's calendar.

Within two months, her weekly "leadership therapy" hours dropped from eight to two. Rachel and David weren't best friends, but they had developed a working rhythm that didn't require Simone's constant presence. More importantly, her directors started resolving friction with each other before it ever reached her.

She hadn't abandoned her team. She'd finally started developing them.

REFLECTION PROMPT

What "fires" in your world actually deserve your attention—and which ones need a better system, not a faster savior?

Life Beyond the Smoke

Reclaiming your time isn't just about becoming a better CTO. It is much more than this. It is about becoming a more sustainable human. The job will take as much as you are willing to give. Without boundaries, it will happily consume your nights, your weekends, and even your relationships. But when you build systems that don't depend on your constant presence, by empowering leaders, negotiating realistic expectations, and stopping yourself from jumping at every flicker of urgency, you gain back something invaluable: time to live your life.

Time for your family. Time for rest. Time to think without a dozen Slack notifications breaking your focus. And paradoxically, the more you protect that personal time, the more effective you become at work. You show up calmer, clearer, and better able to lead without defaulting to adrenaline. Escaping the firefighter cycle is not just a productivity hack. It's also a survival strategy for both the company and you.

Leaving the Firefighter Behind

Your job as the CTO is not to be the ever-present rescuer. It's to align technology with the business, steward your resources, and build the organization that builds the product. Those responsibilities don't change, no matter how chaotic your company becomes.

What must change is how you approach them. If your weeks are defined by endless emergencies, you are not truly leading. You are reacting. And while that might make you feel indispensable, it will eventually burn you out and hold your company back.

Stepping out of the smoke requires courage. It also requires trust in your team, in your processes, and in yourself. The reward, however, is enormous: a company that can scale, a role that truly reflects your value, and a life that no longer feels like it exists only in the margins. The question is not whether the fires will happen. They will.

The question is: Will you keep living in the flames, or will you finally build the firebreaks and step into the clear air of leadership?

PART II

THE CTO AS A MULTIPLIER

CHAPTER 5

The Leverage Revolution: Amplifying Your Impact

True CTOs don't build engines of effort. They build engines of leverage. Leverage, not labor, is your real product.

The first instinct of new CTOs is to measure their value by their own output. How many problems did I solve? How many fires did I put out? How many decisions did I make? It feels satisfying. It feels productive, but it's a trap, because acting as the main point of control is the lowest form of leadership.

Every action you take, every technical decision you personally resolve, every meeting you micromanage is a moment where your company is relying on your effort instead of your system. A CTO's real job isn't to work harder. It is to build *multipliers:* self-managing teams, scalable systems, resilient processes, and people who don't just reduce your workload but who expand your company's capacity to execute without you at the center.

In Chapter 1, you were introduced to the CTO Levels™ Framework, a grid depicting levels 0–10 of the organizational complexity the CTO faces depending on their team size and budget. Five leadership growth areas are needed to navigate the growing complexity identified by the levels in the CTO Levels™ Framework:

- Levels 0 and 1 are where the **individual contributor CTO** can be found. The work still revolves around you. You are still coding, shipping, and defending your worth through personal technical output. Success is measured by what you do yourself, not by what the organization produces without you.
- Levels 2 and 3 are the **decision-maker CTO**, directing traffic, approving architectures, and carrying every hard call. Although not as hands-on as the first level, it is still unsustainable at scale. Here, the business needs you to make decisions, and your time is dominated by tactical problem-solving rather than long-term system building.
- Many CTOs bump their heads when it comes to levels 4 and 5, where you shift into **system builder**: You hire, coach, and empower leaders who make good decisions without your constant oversight. At this stage, the CTO Levels™ Framework emphasizes assessing both the team's actual competency and the company's operating level.
- In levels 6 and 7, you become a **strategic connector**, shaping the company's direction and translating between the executive team and the technology organization. The framework reinforces that influence, not personal output, defines your leverage: You orchestrate alignment across functions and ensure that technology execution reflects business strategy.

- Finally, in levels 8+, you operate as a **multiplication leader**: someone whose presence accelerates the growth and alignment of others, not because you are doing their work but because your influence, clarity, and design make the entire organization smarter, faster, and more resilient. According to the CTO Levels™ grid, your impact here is measured not by individual contributions but by the organization's ability to deliver results independently of you.

The more complex your technology organization, the less your value is tied to personal output, and the more it's tied to how effectively you multiply others' efforts. Leverage, and not effort, is the product you are here to create. Using the CTO Levels™ Framework, leaders can systematically identify gaps, align team capability with company goals, and craft a deliberate growth path that transforms both themselves and their organization.

The CTO Leverage Hierarchy

So, how do you actually multiply? How do you stop measuring your effectiveness by how many things cross your plate, and start building impact that compounds even when you are not in the room? The answer lies in leverage as hierarchy, not tactic. Each layer offers more scale than the one below it, but only if you move up intentionally rather than getting stuck optimizing the lowest rung.

- **Layer 1—Time (linear effort):** This is the most obvious lever and the most fragile. You can restructure your calendar, batch meetings, and protect focus blocks. However, time always scales linearly. You gain an hour by removing an hour, and no amount of discipline creates more. If your impact depends on your presence, you have not created leverage but simply refined your workload.

- **Layer 2—Tools (efficiency):** Above time sits tools. Dashboards, automations, and integrations reduce friction and eliminate manual effort. They can dramatically improve efficiency, but they do not change how decisions are made. Tools accelerate behavior; they do not correct it. Used well, they free capacity. Used poorly, they help organizations move faster in the wrong direction.
- **Layer 3—Code (static automation):** When you automate a recurring process through software, you introduce leverage that can operate continuously and at scale. Code is powerful because it replaces effort with execution. But it remains deterministic. It does exactly what it was designed to do: no more, no less. Code can scale work, but it cannot reason, adapt, or navigate ambiguity.
- **Layer 4—AI (synthetic judgment):** Unlike static code, AI enables synthetic judgment, meaning it can interpret signals, handle edge cases, and make judgment-based decisions that once required human cognition. This is a shift from automating tasks to automating thinking. When designed intentionally, AI becomes a bridge between human judgment and technical execution, allowing systems to adapt without constant intervention.
- **Layer 5—Systems (organizational logic):** Above technology sits something powerful: systems. These are the processes, rituals, and shared mental models that allow organizations to repeatedly make good decisions without heroics. Systems multiply effort and understanding, turn

chaos into rhythm, create consistency with rigidity, and prevent the company from depending on any single individual, including you.

- **Layer 6—People (exponential potential):** This is where leverage becomes exponential. Research into peak performance confirmed what experienced CTOs already know: The greatest impact comes not from what you do directly but from crafting the conditions where others consistently do their best work (Kotler, 2014). In states of deep focus, clarity, and engagement, productivity and creativity multiply far beyond baseline output.

The true measure of a CTO, then, is not the lines of code written or problems personally solved, but how often their teams operate at that high-caliber input level. By designing environments that protect focus, clarify purpose, and remove unnecessary friction, you shift your value from execution to enablement. And when you hire, coach, and empower leaders who can build systems and teams of their own, your leverage compounds. Each leader becomes a multiplication node, extending your impact far beyond your time, tools, or technical output.

This is the hierarchy of leverage. If most of your energy goes into managing time, you remain anchored at the bottom. If you invest in people and systems, you operate at the top, where progress continues even when you step away, disconnect, or think long-term.

The Personal Leverage Ratio

Here is a question I often ask CTOs when I spend time coaching them: "If you vanished for a month, how much of the company's forward motion would actually stall?" The answer is mostly predictable. For many, it's: "Almost everything." Which tells me two things:

1. You are probably overfunctioning in areas that others should own.
2. Your leverage, your ability to create impact beyond your effort, is dangerously low.

To quantify this, we use a concept known as the *personal leverage ratio*. It's not a perfect science, but it offers a gut-check on how much of your work is multiplying versus merely maintaining. Start by tracking your weekly hours and categorizing them into two buckets:

- **Direct output hours:** Time spent on work that would stop if you stopped (coding, approvals, decisions, and meetings that only move forward because you are there).
- **Multiplicative hours:** Time spent on work that builds capacity beyond yourself (coaching leaders, improving systems, aligning executives, designing automation, and creating clarity).

Your leverage ratio = Multiplicative hours ÷ Direct output hours

Most overextended CTOs have ratios under 1.0, meaning that for every hour they spend multiplying, they spend more than that keeping the lights on. Sustainable CTOs aim for 1.0 or higher, where at least half of their time builds enduring capacity. Elite CTOs, especially at scale, get to 2.0 or beyond, where the majority of their hours create impact that compounds long after the meeting ends or the product ships.

Studies of digital work patterns reveal just how fragmented modern knowledge work has become. Harvard Business Review reported findings from a Pegasystems study that analyzed nearly 5 million hours of desktop activity and discovered that employees switch between apps or browser windows roughly 1,200 times per day (Murty et al., 2022). That means the average worker is toggling every 20–40 seconds, far from the traditional multitasking and closer to continuous microswitching. Every moment between a CRM, a messaging tool, a dashboard, or an email client creates a tiny cognitive cost, and when multiplied by hundreds of cycles, these microfractures in attention silently consume a massive share of the workday.

The cognitive overload is not just an inconvenience; it is also a measurable loss of cognitive momentum. Research by Qatalog and Cornell's Ellis Idea Lab shows that it takes about 9 minutes and 30 seconds for a worker to reestablish productive workflow after even a simple context shift like toggling out of a task (Nawrat, 2021). This is distinct from the deeper 23-minute recovery window identified in earlier studies by Dr. Gloria Mark, which reflects the time required to return to sustained deep work after a major interruption (Smith, 2018). Together, these findings paint a clear picture: Modern digital environments are not just distracting; they are also designed in ways that repeatedly eject people from flow, forcing them to rebuild focus dozens of times per day.

In one 7CTOs peer group meeting, we mapped this across real examples. One CTO of a 40-person engineering organization was spending nearly 30 hours per week in decision-making meetings, approvals, and escalations. His leverage ratio? 0.3. He was busy, but his company was fragile: Every decision bottlenecked at his desk. After six months of deliberate coaching, delegation, and system building, he flipped that ratio to 1.2. His calendar didn't get lighter, but his impact grew exponentially because his hours were now shaping leaders and systems, not just output.

The Replacement Test

There is a simple way to test whether you are truly building leverage or just maintaining motion. Ask yourself, *Could my organization survive, or even thrive, without me for a month?* This isn't about making yourself irrelevant. It's about ensuring that your value lies in the systems and leaders you have built, not in your personal availability.

If the honest answer is "no," that is not a failure; that is feedback. It means you are still too enmeshed in operational gravity. Maybe your directors are not fully empowered to make calls without you. Maybe your processes aren't resilient enough to handle change. Maybe your peers or CEO still see you as the executor-in-chief rather than the architect of a self-sustaining machine.

The replacement test is not meant to diminish your importance. It is meant to surface the areas where you are overfunctioning so you can begin to transfer ownership, fortifying systems and building leaders who make the organization stronger even in your absence.

In practice, this is used as a diagnostic tool. We note the functions or outcomes that would wobble without you. Maybe architecture reviews would grind to a halt. Maybe hiring would stall. Maybe Product and Engineering alignment would fray. Each of those becomes a focus area for the months ahead, a target for coaching, process, or structural change.

Because a CTO who can step away without the company collapsing is not a less valuable leader. They are a *more valuable one.* You will become a leader whose legacy is a resilient, scalable organization, not a fragile one propped up by your constant presence.

REFLECTION PROMPT

If you vanished for a month, what would actually stop? What would keep moving? And which of those gaps is your next opportunity to build leverage?

Refactor in Action

Priya was proud of how much she could carry. Her calendar was a blur of meetings—approvals, architecture reviews, and cross-functional escalations. Her notifications pinged every few minutes with some version of, "Can you decide...?"

It felt good. The CEO praised her responsiveness. Product leaned on her to settle debates. Engineers lined up for her judgment calls. Priya was the hub, and the wheel spun fast.

Until it didn't.

One Thursday night, after 12 hours of context switching, she opened the scalability plan the investors were waiting on. Blank. She stared at the blank page until her vision blurred, then closed the laptop and whispered to herself, "I've got nothing left."

The next morning, another escalation landed. She almost replied, but paused, then forwarded it to her Director of Engineering: "You own this. I trust your call."

Then she braced herself.

By evening, the problem was solved. Not perfectly, but without her.

Over the next month, she tracked her time: work that only she could do versus work that others could own. The imbalance was brutal. Most of her energy went into output, not leverage.

She started making changes: promoting managers, creating a biweekly forum for architecture debates, blocking strategy time, and refusing to give it back.

Not everything clicked. Some handoffs got messy, some decisions slower. But Priya began to see the truth: The more she held on, the weaker the system became. Letting go didn't shrink her role. It multiplied it.

Leverage Is the Product

The companies that grow fastest, and the CTOs who thrive longest, share a mindset: *Leverage is the product.* The code will change, and the tools will evolve. The organization chart will shift. But the only thing that endures is the capacity you build: the people and systems that carry your influence forward even when you are not in the room. True CTOs do not just ship features; they also ship leverage. This capacity is what allows you to not only grow the company but to reclaim your life, your energy, and the clarity that first drew you to leadership.

The Mathematics of Impact

This isn't just a leadership philosophy; it is a mathematical law of the office. Andy Grove (2015), the legendary CEO of Intel, codified this in his "Managerial Output" formula:

$$Output\ (CTO) = Output\ (Org) + Output\ (Neighboring\ Orgs)$$

Notice what is missing from that equation: *your individual output.* According to Grove, your personal "productivity"—the code you write, the emails you clear—is mathematically irrelevant to your value unless it serves as a high-leverage activity that multiplies the activities of the teams around you (Grove & Horowitz, 2015). Your productivity is defined as:

$$Productivity = \Sigma(L \times A) \div Time$$

Where:

- L is the activity's leverage
- A is the activity being done
- T is the time spent

The "High Output" CTO understands that their job is to *maximize the leverage of their activities,* not the *volume of their tasks.* If your presence is required for a system to function, you haven't increased the capacity; you've simply become a bottleneck in the calculation.

The Leverage Audit: Addition vs. Multiplication

To move from a "summing" mindset to a "multiplying" one, you must audit where your energy goes. Are you adding tasks or multiplying potential?

The personal leverage ratio gives you a number, but a number alone does not tell you where to act. The Leverage Audit is the diagnostic layer underneath it. It takes the abstract question—*Am I multiplying or just maintaining?*—and turns it into a concrete, week-by-week exercise you can repeat any time your calendar starts to feel heavy again.

Step 1: Capture

For one full week, log every meaningful activity you engage in. I am not talking about every Slack message or bathroom break. I mean every block of time where you are doing work: reviewing code, sitting in a meeting, making a hiring call, resolving an escalation, writing a spec, coaching a lead, unblocking a team. You are building a raw inventory of where your hours actually go—not where you think they go. Most CTOs who do this for the first time are genuinely surprised by what they find. The calendar lies. The time log does not.

Step 2: Classify

Once you have the inventory, take each activity and ask two questions:

- **First:** *Am I adding, or am I multiplying?* Adding means you solved a problem once, for one situation, and the effort ended with you. Multiplying means you changed the conditions so that the problem gets solved repeatedly, by others, without your involvement.
- **Second:** *What happens to this work after I am done?* If the honest answer is "Nothing—it is finished, and nothing changes systemically," that activity belongs in the low-leverage column. If the answer is "This changes how problems get solved going forward," it belongs in the high-leverage column.

Be ruthless with yourself here. Most CTOs who complete this step discover that 70%–80% of their week lives in the low-leverage column. That is not a failure—it is a diagnosis. You cannot fix what you have not named.

Step 3: Convert

This is where the audit becomes actionable. For every low-leverage activity on your list, ask: What would the high-leverage version of this look like, and what is the multiplication effect?

The multiplication effect is the downstream impact of converting a single instance of effort into a systemic change. It is the reason you are doing the conversion in the first place—not just to free up your time, but to expand the organization's capacity to operate without you in the loop.

Each conversion becomes a concrete initiative you can prioritize in the weeks ahead. Some will be quick wins—writing a piece of documentation, delegating a recurring approval. Others will be multimonth investments—building a decision rights framework, coaching a director into full ownership of architecture reviews. The point is not to convert everything at once. The point is to see the full landscape and choose deliberately.

To make this tangible, here is what a completed audit looks like across three common activity types:

Activity type	Low leverage	High leverage	The "multiplication effect"
Problem-solving	Patching a critical bug yourself.	Leading a blameless postmortem to fix the systemic root cause.	Prevents the bug from ever recurring for 50+ engineers.
Communication	Answering the same tech question in Slack 10 times.	Creating "golden path" documentation or an ADR.	Provides "on-demand" clarity for every current and future hire.
Decision-making	Being the final "green light" for every public relations campaign or hire.	Defining decision rights and cultural principles.	Empowers the team to make 1,000 correct decisions without you.

These are examples, not an exhaustive list. Your audit will surface activities specific to your context—vendor negotiations you should not be running, sprint ceremonies you should not be attending, technical decisions you should not be approving. The exercise works because it forces you to confront a simple truth: Every hour you spend adding is an hour you are not spending multiplying. And over time, that imbalance is what separates the CTO who scales from the CTO who stalls.

The Paradox of the Constraint

When you stop measuring your value by how much you do and start multiplying others' impact, the company begins to feel lighter, faster, and more capable. You are no longer holding the wheel on every decision, nor are you burning every weekend fighting fires. Your leaders rise. Your systems stabilize. You finally feel like you are steering the ship, not bailing water.

However, there is a paradox here that trips up even the most seasoned CTOs: As you climb the hierarchy of leverage, you will find yourself staring at the uncomfortable truth that *you are now the biggest constraint.*

Not because you are slow or incapable, but because old habits—holding decisions, answering every question, and inserting yourself into every priority—start to drag the very machine you have built. The more you grip the controls, the more friction you create. And the more friction you create, the more slowly everything moves, including your growth.

Loosening the Grip

To fully step into the role your company needs now, you cannot just build leverage; you also have to get out of its way. You must shift from being the central processor to being the catalyst.

In the next chapter, we will explore how to find the true bottlenecks in your organization (including yourself), engineer decision rights so work flows without constant escalation, and master the art of coaching, questioning, and disappearing. The next evolution of your leadership is not about holding more tightly—it is about loosening your grip and discovering that, when you do, everything starts to move faster.

CHAPTER 6

Bottleneck to Catalyst: Accelerating Through Letting Go

The more you let go, the faster everything moves—including your own growth.

In every system, there is a bottleneck: the single factor that dictates how fast everything can move. In code, it might be a fragile dependency. In manufacturing, a lone machine that can't keep pace. In your tech organization? More often than we like to admit, the choke point is the person wearing the CTO badge.

I have been that bottleneck. Early in my career, I took pride in being "in the loop" on everything. Every architecture review, every priority debate, every hire crossed my desk. My fingerprints were on every big decision, and I wore that as a badge of honor.

Until one day, a CTO friend of mine listened to me rattle off my schedule and said, "You realize you are the single point of failure, right?" It stung mostly because it was true. My team wasn't slow. My processes weren't broken. *I was the throughput limit.*

And I am not alone. In every 7CTOs coaching group, we see the same dynamic. CTOs start decisive, involved, and highly effective. But as the organization scales, that same effectiveness turns into a drag. Decisions stack up in your inbox. Everyone stops to "loop you in." You feel needed, but the company is moving more slowly than it should.

The first step to breaking this pattern is awareness. That is where the *constraint map* comes in. Pick a deliverable—a product launch, a big refactor, or a new vendor rollout—and trace it from start to finish. Note every handoff, every meeting, every decision point. Where does the flow slow down?

Most CTOs are startled by what they find. The friction isn't buried in code or process. It's sitting squarely on their calendar.

Decision Rights Engineering

Once you identify the bottleneck, the solution is not to work longer hours or respond to Slack faster. It's to make the decision pathways explicit so you are no longer the default decider.

In one coaching session, a CTO of a 120-person team admitted he was still approving *early* vendor contracts and personally greenlighting all backend architecture changes. Not because he did not trust his leaders, but because no one had ever drawn a map of who could decide what. In the absence of clarity, everyone went to the safest place: him.

I built a simple *Decision Rights Matrix*. For each recurring decision type, from architecture choices to budget approvals to feature prioritization, we defined

- *who owns it outright* (even if that is not you, the CTO).
- *who must be consulted.*
- *who just needs to be informed.*
- *which decisions truly require the CTO's intervention.*

Within two months, his volume of escalations dropped by half. His leaders felt empowered, not abandoned. And he stopped being the human router for every workflow in the company. Clear decisions are not bureaucracy; they are velocity.

The Science of Flow: Westrum and DORA

The DORA research reinforces a pattern that many seasoned CTOs learned the hard way: Elite technical performance is not the result of better code alone; it is the result of decentralized authority. Across a decade of data, high-performing organizations consistently show higher deployment frequency and lower lead times when decision-making is clear, distributed, and trusted. In other words, the system keeps moving even when you are unavailable, which is exactly the point.

When you remove yourself as the manual "green light," you are not stepping back from leadership; you are rewiring the organization. You are shifting how information flows, how risk is handled, and where authority lives. Sociologically, this movement tracks closely with the Westrum model of organizational culture, which measures whether a system is oriented around power, rules, or performance. It defines three cultural patterns: pathological, bureaucratic, and generative.

Many CTOs I work with discover they are operating in the middle of the Westrum model. Not overtly political or fear-driven, but bureaucratic. Authority follows title, and information moves through approved channels. Progress slows the moment a key decision-maker disappears. In that state, the organization only functions while you are present.

The goal is not perfection in the organization; it is to move toward a generative culture, where information is actively shared, failure becomes learning, and authority is pushed to the edge where the work actually happens. This is not about being "nice." Rather, it is about building a system that can self-correct and scale without constant intervention. This is leverage in its purest form.

Of course, letting go is not a binary switch. You cannot move from being a micromanager to being hands-off overnight and expect the organization to flourish. Letting go is a craft, and like any craft, it needs structure.

You coach when your team does not need an answer, but they need your thinking reflected back to them. This is where catalytic questions matter more than solutions. "What would you need to be true for this to work?" "What risks are we not naming?" The goal is not to solve the problem yourself but to expand how your leaders think, so fewer problems escalate to you in the first place.

You decide when the stakes are high or when alignment is missing. This is not about last-minute vetoes. High-leverage decisions clarify the system, not just the task. When you engage early, your decision becomes a North Star rather than a surprise obstacle.

And then there is releasing, the hardest and most important move of all. You step back when your presence would only slow things down. This does not mean withdrawing or disengaging. It means trusting your leaders to operate within the guardrails you have already built.

I'll admit that releasing was the hardest for me. The first time I skipped a major road map review, I felt a sharp pang of irrelevance. But the team delivered flawlessly. That moment taught me an essential distinction: They did not need me to hold the pen; they needed me to hold the vision.

Still, releasing does not mean checking out. There is a dangerous middle ground where senior leaders detach so completely that they drift into becoming "PowerPoint CTOs," leaders whose decisions are abstracted from the reality of the codebase and the systems they oversee.

This is where the Deep Dive Audit becomes critical. Once a month, choose a narrow slice of the system, read the code, run the build, and trace the logs. The idea is to follow the details all the way down. The goal is not to fix anything yourself but to audit the system's health and verify that your trust in the organization is warranted.

Done consistently, the Deep Dive Audit reinforces two truths at once: Your leaders are trusted to drive, and you remain close enough to the critical systems to lead with credibility and authority. As you shift from deciding to coaching and, when appropriate, disappearing, your leaders grow, the system accelerates, and you finally reclaim the space to focus on the work that only you can do.

Question Leadership

Old-school leadership teaches us to provide answers. It feels efficient. Someone's stuck; you swoop in with the fix. Problem solved. But over time, this approach makes you the bottleneck twice over: once because everyone waits for your answers, and again because no one learns to think without you. The alternative is Question Leadership. Instead of telling your team what to do, you help them sharpen their own judgment through inquiry.

One CTO in a recent 7CTOs peer group meeting shared how he transformed his weekly check-ins. Instead of dictating priorities, he started each session with questions: "What is blocking you that I can unblock?" "What trade-offs are you wrestling with?" "How does this decision align with our North Star metrics?"

The shift was subtle but profound. His team began anticipating these questions, arriving prepared with their own solutions instead of problems. Over time, they needed him less, not because he was less available but because they were more capable of dealing with issues without the CTO's presence. Obviously, this is not a trick to dodge responsibility. There are moments where clear direction is essential. But if every interaction ends with you answering, you are not leading; you are just scaling dependency.

REFLECTION PROMPT

Where are you acting as the constraint today? Which decision, habit, or responsibility could you release—with clarity and support—to move your team, your company, and your own growth forward?

The Ownership Culture

Delegation solves today's problems. Ownership builds tomorrow's. A true ownership culture is not one where people just do what they are told. It is one where they *think like owners*: understanding the company's objectives, making decisions that fit the context, and acting without waiting for a rubber stamp.

Research on team autonomy continues to validate this principle. Spotify's squad model, for example, showed how small cross-functional teams of 6–12 people can operate as independent, high-velocity units when given both autonomy and clear alignment (Šmite et al., 2023). Their "high autonomy, high alignment" philosophy inspired organizations around the world, yet many failed when they attempted to replicate the structure without the underlying cultural and architectural foundations.

Recent studies, including large-scale analyses of autonomy in software teams, confirm this gap: Autonomy only drives performance when paired with deliberate design, shared norms, and cohesive technical ecosystems (Šmite et al., 2023). This lesson is consistent across the research: Ownership is not granted by reorganizing teams on a chart. It is earned through intentional systems that enable teams to truly act like mini-start-ups rather than departments with new labels.

This does not happen by accident. It starts with *context*: sharing the "why" behind the decisions, not just the "what." It thrives on *clarity*: giving teams the boundaries within which they can move fast. And finally, it grows through visibility: celebrating wins, analyzing losses, and sharing lessons so people feel supported in taking initiative.

When your team operates this way, your role dramatically changes. You are still guiding strategy and building systems, but you are no longer the approval gate for every task. Your presence becomes a force multiplier, not a speed bump.

The Recovery Plan: When Autonomy Stumbles

Well, let's be honest: Sometimes, autonomy backfires. A leader makes a call that tanks a sprint. A product pivot drifts too far from the company's strategy. A technical choice triggers unexpected pain. The instinct, for many CTOs, is to take the reins back: "Clearly, I cannot let this happen again." But every time that you do that, you teach your team that ownership is conditional—it only lasts until the first misstep.

A better approach is to have a *recovery plan* ready. When something goes sideways

1. **Start with inquiry, not blame:** Ask, "What did you see at the time? What context were you missing? What would help next time?"
2. **Fix the system, not just the outcome:** Was this a judgment issue, a gap in clarity, or a missing safeguard? Patch the weaknesses so autonomy improves, not shrinks.
3. **Reaffirm trust:** Make it clear that one mistake does not mean a permanent rollback to micromanagement.

One time, a CTO shared how a badly timed tech choice cost them three weeks. Instead of stepping back into the role of the sole decision-maker, she worked with her lead to document a clear architecture playbook and instituted a lightweight peer-review step for major tech bets. The result? The leader grew, the system improved, and she did not reclaim the bottleneck role.

Refactor in Action

Mark's wife had begged him to leave his laptop at home. "Just one week. The team can handle it."

He wanted to believe her. They booked flights, packed bags, and, for the first time in years, he told his team, "I'll be offline."

Day two of the trip, his phone buzzed on the nightstand. Then again. By breakfast, Slack was a wall of red dots. Vendor questions. Architecture debates. Hiring approvals. Every thread ended with the same line: "We're waiting on Mark."

He tried ignoring it. By the third day, he was taking calls from the hotel balcony while his family swam in the pool. By day five, he had given up the pretense—his laptop was open full-time, the vacation just a change of scenery.

When he returned, his CEO was blunt. "You've built a system that collapses the moment you step away. That's not leadership, Mark. That's a liability."

The next week, Mark gathered his directors. "If I vanish for a week, the company should keep moving." Together, they mapped decision rights—what they owned outright, what they'd consult him on, and what only he would decide. It wasn't clean at first. Mistakes were made. But gradually, the weight shifted.

From Bottleneck to Catalyst

This shift from the person who holds everything to the person who unlocks everything is not easy. It requires self-awareness, courage, and a willingness to redefine what leadership looks like. It means confronting a deep-seated belief most of us hold as we rise through the ranks: that our value is measured by our presence, by our fingerprints on every decision and deliverable.

But when you make the shift, everything changes. Your team stops waiting for you. Decisions flow without the bottleneck of your calendar. The company regains its rhythm. And you finally get to focus on the work only you can do: shaping the company's long-term vision, aligning your fellow executives, and architecting systems and leaders that scale far beyond your daily hours.

I have seen this transformation in myself, and I see it every quarter in the CTOs we coach through 7CTOs and Coach EXE. The pattern is nearly universal: The ones who cling to control, convinced this constant presence is synonymous with value, eventually burn out. Their companies stall, slowed by the very person who was once the engine. The ones who intentionally let go, with clarity and support, discover that their teams rise to meet the challenge. Their influence deepens, their careers accelerate, and their companies begin to run at a speed and stability that was previously impossible when everything revolved around one person.

And here is the paradox: The more you release, the more everything, including you, accelerates. You don't fade into irrelevance; you graduate to a new level of leadership, where your value is measured not by how much you hold together but by how well you've built a machine that doesn't need holding at all.

But letting go cannot be just a mindset. If you want your absence to create momentum rather than chaos, you need to design for it. You need a system that doesn't just follow delegation but makes delegation part of the company's operating model. A system that builds autonomy progressively, creates tight feedback loops, and makes your presence optional, so that when you *are* in the room, it's by choice, not necessity.

The next chapter will explore exactly that: how to build those systems. We will map out a Delegation Canvas to match the right work to the right people. We will talk about designing roles and processes that thrive without your direct input, and how to use progressive autonomy to grow leaders rather than overwhelm them. And we will break down a framework for those of us who, despite our best intentions, still struggle to fully let go.

Becoming a catalyst is not about stepping back. It's about building a company that is designed to run (and win) even when you are not there.

CHAPTER 7

Delegation as Design: Creating Systems That Don't Need You

The true test of your leadership is not how
the company runs when you are there.
It is how it thrives when you are not.

Most CTOs think of delegation as a simple act: hand off tasks so that you can focus on "more important things." You pass along the software development life cycle, the vendor reviews, maybe even some hiring calls, and you breathe easier, at least for a moment. But real delegation, the kind that actually scales you and your organization, is not about shuffling work off your plate. It is about designing a system that can run without you as the driver. A system where your presence is valuable but not essential. Where teams know what to do, how to do it, and when to bring you in—and, just as importantly, where you know when you can step away without losing momentum.

This is where the Westrum Organizational Continuum becomes an invaluable mirror. Developed by sociologist Dr. Ron Westrum through research in safety-critical fields like aviation and medicine, the model explains how information and authority really move through an organization. Each is defined by how teams respond to failure, risk, and bad news. In

- **pathological systems**, information is hoarded, and messengers are punished.
- **bureaucratic systems**, it is trapped in a process.
- **generative cultures**, information is actively sought, messengers are trained, and risks are shared.

For a CTO, this framework shifts the question of delegation from "What can I give away?" to "What kind of environment am I building?" It highlights the difference between being Chief Approver, where progress awaits your consent, and becoming an Architect of Environment, where trust, clarity, and psychological safety allow technical leverage to scale. Delegation, seen this way, is no longer a productivity tactic; it is a cultural design choice.

It is how you build an organization that survives and thrives in your absence. In this chapter, you will explore how to do exactly that. You will use a *Delegation Canvas* to match the right work to the right people. We will discuss how to *design for absence* so your systems are not propped up by your availability. You will learn about *progressive autonomy* so your leaders grow without drowning. Finally, you will learn how to create *feedback loops* that turn delegation into a learning engine rather than a gamble.

If Chapter 6 was about letting go of the bottleneck, this chapter is about making sure that what you have built keeps accelerating when your hands are not on the wheel.

The Delegation Canvas

Bureaucracy often occurs when the person with the most information (the engineer) has the least authority, and the person with the most authority (you) has the least information. We either dump tasks wholesale, overwhelming a leader who is not ready, or we hoard high-leverage responsibilities because we are convinced that no one else can do them. Both approaches backfire.

That is why I use the *Delegation Canvas,* a simple tool we developed in my coaching experience, to match work to people based on two dimensions: the *impact* of work and the *readiness* of the person.

On one axis, plot the work:

- **High leverage** (decisions or systems that shape the organization long-term) versus **low leverage** (tasks that sustain operations but don't change the trajectory).

On the other, plot the person:

- **High readiness** (they have the skill, context, and confidence) versus **low readiness** (they are still learning or lack context).

It should look like this:

Work/Person	High readiness	Low readiness
High leverage	This is where the future leaders will be built. You should give them the keys and let them drive, checking in only for alignment.	Requires codelegation, meaning that you stay involved, but not as the *doer*—as a *coach*, transferring judgment as you go.

Work/Person	High readiness	Low readiness
Low leverage	This work can be fully delegated. Let your leaders own it, freeing you for bigger things.	Tasks in this quadrant should be automated or eliminated outright.

This canvas does more than assign tasks. It reveals your real job: not to do the work, but to grow the people who can own the highest-leverage work without you. Over time, as their readiness increases, your hands-on involvement should steadily decline.

Absence of Talent

The talent bridge is the uncomfortable territory every CTO eventually confronts: *What if the team simply is not ready?* Many leaders look at their Delegation Canvas and realize their organization sits squarely in the "low readiness" quadrant. Handing over meaningful ownership feels reckless, especially since in most cases (almost guaranteed), it will end in failure. This fear is what creates the competence vacuum: the single most common reason CTOs fall back into firefighting. When no one else seems capable, doing everything yourself feels safer, faster, and even responsible. But this instinct traps you in operational gravity and prevents your team from ever developing the capability you need them to have.

The key is remembering that you cannot delegate execution to someone who lacks the skill. However, you *can* delegate context. Instead of abandoning junior teams or shouldering all the work, you can move into a deliberate coaching posture. Narrate your thinking instead of quietly solving the problem. Record a five-minute Loom walking through why you inspected certain logs, evaluated specific failure modes, or ruled out particular architectural paths. Require your team to write the *first* draft of decisions or technical specs, even if you still provide the final review. This approach builds judgment without placing the system at risk, accelerating competence through structured exposure through your reasoning.

If you still find the gap too wide to bridge with coaching alone, the solution shifts from technical involvement to strategic hiring. If your team is not yet ready to take the keys, then your highest-leverage priority is finding the senior lieutenant who *can*. Hiring someone who can absorb entire categories of decisions is not a luxury; it is the infrastructure that enables you to rise into true executive leadership. In the absence of this hire, you remain the bottleneck. With it, you unlock the capacity for scale, resilience, and the distributed ownership your organization needs to grow.

Design for Release

Designing for release means creating systems where your role is additive, not structural. Netflix's "freedom and responsibility" culture, eliminating vacation policies and formal hierarchies while maintaining high performance, proves that radical autonomy works at scale. Their culture deck, viewed over 5 million times and called "the most important document ever to come out of Silicon Valley" by Sheryl Sandberg, demonstrates intentional absence-design (Shontell, 2013).

It means roles, processes, and decisions are so clearly defined that the machine hums even when you are not around. Start by looking at where your organization still depends on your presence rather than your *design:*

- Are there decisions only you can approve because no one knows the criteria?
- Are there processes that stall if you are not in the meeting?
- Are there teams that wait for your feedback rather than owning their outcomes?

Then, codify the rules of the system:

- clear decision rights (who owns what)
- documented guardrails for quality and risk
- rituals like weekly syncs or decision logs, so context is shared even when you are not the one sharing it.

One CTO I coach put this to the ultimate test by taking a four-week sabbatical. Before leaving, she documented decision matrices, clarified escalation paths, and empowered her directors to resolve anything short of a security breach without her. The result? The team didn't just survive. They produced excellent results. Velocity increased because fewer things bottlenecked on her. When she returned, she found an organization that no longer needed her as a daily governor. She could finally act as the strategic architect.

Progressive Autonomy: Building Capability Without Breaking People

One reason delegation fails is that we treat it as binary: Either we do the work ourselves, or we hand it off completely. For leaders who are still developing, both extremes are risky. Too much hand-holding stunts their growth. Too much autonomy, too soon, can overwhelm them and trigger costly mistakes.

The solution is *progressive autonomy*, which translates to building capacity through calibrated challenges. Think of it as teaching someone how to swim: You don't throw them in the deep end, but you also don't keep holding the back of their float forever.

Here is how it works:

1. **Start with shadowing.** Let them watch you handle a situation, narrating your thinking process.
2. **Move to co-decision-making.** They lead the discussion; you provide feedback and final approval.
3. **Shift to full ownership with oversight.** They make the call, and you review the outcome after the fact.
4. **End with trust and audit.** They own the domain outright; you only step in for periodic reviews or when there is a significant change in context.

This staged approach allows leaders to work faster than micromanagement, but with less risk than total abandonment. Over time, more and more of your organization's critical functions land in Stage 4, and your bandwidth for the true CTO work multiplies.

The Feedback Loop: Turning Delegation Into a Learning Machine

Even the best-designed delegation system will fail if it's static. Without feedback, you cannot tell whether your team is growing, if your processes are working, or if your absence is being productive. Delegation without feedback is abdication. Leaders take on new responsibilities, mistakes happen, and no one learns. The CTO either swoops back in or silently loses confidence, and the cycle repeats.

That is why every delegation system needs a feedback loop. Drawing from Peter Senge's *Fifth Discipline* (a core reference in my bookshelf), feedback is not just about identifying errors; it's also about creating learning organizations where every handoff becomes an opportunity to improve. Research validates Senge's framework: Organizations with systematic feedback loops show measurably better outcomes.

Practical feedback loops include

- **Decision retrospectives:** After major calls, discuss what context was clear, what signals were missed, and what adjustments would make the next decision smoother.
- **Metrics with meaning:** Use key performance indicators that reflect system health (velocity, incident rates, or cross-team dependencies), not just output, to see whether autonomy is delivering results.
- **Reflective 1:1s:** Shift check-ins from status updates to capability-building: "What felt heavy this month? What patterns are you noticing? What would help you handle this with less friction next time?"

In one peer group, a CTO implemented a "Friday feedback hour" where each director shared one decision they were proud of and one they wished they had handled differently. Over time, this ritual normalized learning and reduced the CTO's impulse to micromanage, because visibility was built into the culture rather than dependent on his oversight.

The Letting Go Framework

Even with the right tools, CTOs will emotionally struggle to step back. Delegation feels risky; what if something breaks? What if the CEO thinks you are disengaged? What if you stop being "needed?" When I coach, I like to help leaders work through this with the *Letting Go Framework*, a set of practices to reframe control. It consists of five steps:

1. **Name the fear:** Most anxiety shrinks when spoken aloud. Are you worried about reputation, mistakes, or momentum? Clarify it.
2. **Run controlled experiments:** Don't start with a three-month sabbatical. Skip one meeting. Let a director own one decision. Observe; don't interfere.
3. **Measure outcomes, not touchpoints:** Success is about results, not how many approvals you give.
4. **Invest in trust before release:** Train, align, and set expectations so your step back feels like confidence, not abandonment.
5. **Document your unique lane:** Make explicit what only you can do (executive alignment, budget stewardship, vision), so both you and your team know where your focus belongs.

Most leaders find that after a few cycles, their fear subsides. Not because delegation becomes risk-free, but because it becomes *predictable* and *reversible*. You learn to calibrate without reverting to control.

Refactor in Action

"Just handle it," Daniel said, sliding the proposal across the table. His new Engineering manager nodded quickly. Too quickly.

Two weeks later, the proposal boomeranged back in flames: over budget, timelines scrambled, Product fuming. Daniel's calendar was suddenly full with damage control.

This wasn't delegation. This was abdication. He had handed off the work without giving context, guardrails, or criteria. The manager hadn't failed—Daniel had.

That night, tired and frustrated, he rewrote the proposal himself. One thought stopped him: *If I keep doing this, I'll never get out of the loop.*

The next morning, he tried again. This time, he sat down with the manager. They mapped out what success looked like, the budget boundaries, and when to check back in. The manager left with clarity, not just responsibility.

The next proposal came back better. Daniel held the pages longer than he needed to, fighting the urge to mark them up. For once, he didn't. With light feedback, he signed off—and let it go.

Designing for Your Release

Delegation is not about disappearing but about becoming dependable when, and if, needed. The moment your organization runs smoothly without you is not the moment you have become irrelevant. It is the moment you have arrived as a systems-level leader. Because the real measure of your leadership is not what happens when you are present. It is what happens when you are not. Can your team navigate ambiguity, make wise trade-offs, and deliver meaningful outcomes without waiting for your approval? That is the test. And the highest compliment.

When you delegate by design, with intention, structure, and rhythm, you free yourself from the gravitational pull of operational urgency. You create time for strategic thinking. You unlock your ability to invest in vision, relationships, and culture. But perhaps most importantly, you start building the conditions where other leaders grow outside your shadow and stand in their own right.

Still, systems are not enough. You can delegate perfectly, precisely document decisions, and architect clear lanes of ownership. However, if the culture underneath it all is passive, fearful, or dependent, it will not hold. The machine will keep looking for you. And that brings us to the next layer of your leadership: self-managing teams. Because if delegation is about what you hand off, culture is about what they pick up and how they carry it forward without asking for permission every time.

In the next chapter, we will go deeper. You will explore what it really takes to build a team that does not just function without you but also *excels* without you. A team that manages itself with clarity, accountability, and momentum. A team that becomes your ultimate product as a CTO and also your proudest achievement.

Because your calendar is not the bottleneck anymore. No, it's the company culture. And we need to start designing that next.

CHAPTER 8

Self-Managing Teams: Your Ultimate Creation

Modern technology companies do not scale like machines; they scale like ecosystems.

Different studies on organizational design show that the most effective "processing unit" in a company is the autonomous, self-managing team, not individuals. When authority is pushed closer to where information lives, productivity and engagement rise because the latency of permission disappears. Hierarchies slow decisions and degrade context, while adaptive systems distribute judgment and move faster when conditions change.

For a CTO, this represents the highest form of leverage. The role shifts from central processor to architect of the environment. Rather than allocating work or resolving every trade-off, you design the alignment, guardrails, and mental models that allow teams to govern themselves. In this model, flow becomes the default, progress no longer depends on your availability, and your success is measured by the quality and frequency of decisions made without you.

During my coaching sessions, we guide technology leaders to see themselves as architects not of code but of culture and autonomy. This is exactly what this next stage is about: building self-managing teams that make your presence optional and your absence a stronger test of the systems you have built.

If you have ever watched a well-functioning orchestra, you will notice it does not center only on the conductor. Instead, every subgroup knows its role, interacts via clear signals, and adjusts in real time without asking for permission. That is what self-managing teams look like: independent units with a shared rhythm.

Team Topologies for Independence

When dependencies drag down execution, and everything circles back to the CTO, it is not bad luck; it's topology. The structure you create for your team determines how your work moves or stalls. When teams are overly independent, the CTO becomes the "human router": the connective tissue required just to keep the progress flowing. No amount of personal efficiency can compensate for a system designed this way.

This is why team topologies matter: By designing deliberate boundaries with small, decoupled teams aligned to clear domains, you reduce the cognitive load and eliminate unnecessary "bus traffic" (the constant negotiation that slows everything down). Stream-aligned teams can focus on delivering value, while platform and enabling teams support them without becoming bottlenecks.

In this model, Conway's Law works in your favor: When communication paths are clean and boundaries are intentional, the software architecture naturally becomes more modular and resilient (Wokittel, 2023). Team design, then, is not an organizational concern separate from Engineering; it is one of the highest-leverage forms of technical design available to a CTO.

Moving Toward the "Teal" Paradigm

In my coaching, we pair this structural clarity with a self-managing team maturity model inspired by *Reinventing Organizations*. In a Teal organization, the goal is not to eliminate hierarchy, but to transform it. Control gives way to distributed authority, and the CTO shifts from directing work to shaping purpose and limits. Instead of managing people directly, you design an environment where shared purpose guides decisions. When team topologies are combined with this level of autonomy, the organization begins to behave like a living system capable of sensing, adapting, and evolving without waiting for permission from the top.

Stage	Description
Stage 1	Teams rely on centralized decisions (CTO or director). Context drains out, growing fatigue issues.
Stage 2	Teams have graduated autonomy, including some decision rights, and are still awaiting final sign-offs.
Stage 3	Teams make most decisions with guardrails, escalate only strategic risks, and regularly codesign systems.
Stage 4	Teams are near independent: their weekly operations run without constant escalation. They proactively sync with other teams.
Stage 5	Fully self-managing, cross-functional, with continuous improvement built into their structure. No "team silos," only purposeful domains with robust communication.

Every stage increases resilience, speed, and trust. Your goal is not to operate at Level 5 in every function overnight. It is to make progress over time, measuring how dependencies shrink and autonomy grows.

Information Architecture

Teams cannot own what they can't see. That is where *information architecture* becomes mission-critical. For self-management to work, context must be transparent and accessible. At many early leadership stages, significant technical and business data is siloed, meaning it is visible only to you or a few people. Architecture decisions hide in chat threads. Incident data lives only in private dashboards. Road maps are guarded documents.

Design for transparency instead:

- Use public dashboards that automatically update (incident count, lead time, and deployment frequency).
- Maintain an architecture decisions log: a single source document where major engineering decisions are recorded with rationale and context.
- Rotate "content curator" roles across teams: someone who ensures that new information gets surfaced and shared regularly.
- Use open retrospectives, where failures and wins get documented, shared across teams, and baked into culture.

This is not about surveillance. It is about shared awareness. When every team sees what drives decisions and has access to signals, dependency on your explanation disappears. In mature organizations, critical decisions, incident histories, architectural patterns, and delivery metrics are centrally documented, automatically updated, and accessible across teams. Information lives in the space everyone inhabits, not locked behind tickets or private Slack archives.

A fellow CTO recently told me about a flow that she had established in her organization without making it a chore. Each week, she receives "what we shipped, what went wrong, and what we learned" documents from each team. No summary needed. She gets into the autonomy flow, she is still informed, and it doesn't become something that bottlenecks or takes up much of her time.

Decision Frameworks

Self-managing teams can only function well if there is clarity about how decisions are made. Without clear guidance, autonomy can quickly drift into confusion. The very trust that empowers teams can backfire when expectations are not well defined. And worse, when they are assumed but never discussed.

It's not about enforcing rigid control but about offering a decision architecture that gives people the confidence to act. One mental model we use is the idea of **D**ecide, **A**lign, **E**scalate (DAE), which is more of a rhythm than a rule book. In its simplest form:

- If a decision clearly falls within your team's domain and doesn't meaningfully impact others, make it.
- If it might intersect with another team's scope or affect a shared dependency, align first.
- If it involves high-risk implications, such as legal exposure, substantial cost, or a fundamental shift in architecture, escalate it to leadership.

This model works best when teams have agreed-upon thresholds and know where their authority begins and ends. The framework doesn't exist to slow them down. In fact, it is the opposite. It gives them the confidence to act without second-guessing or constantly seeking permission.

REFLECTION PROMPT

What system or behavior still depends on you—not because it should, but because it always has? What would it take to remove yourself from that loop, and how would your team grow if you did?

I've seen this pattern show up repeatedly, just in different forms. Teams with strong intent and capability stall because they are unsure whether they are *allowed* to act. The hesitation is not rooted in incompetence; it is rooted in unclear authority.

In one case, a team tasked with using Claude Code to migrate a monolithic app to microservices became paralyzed by the fear of making the wrong architectural call. Every pull request lingered for days, waiting for the CTO's blessing, because the perceived blast radius felt existential. What changed everything was not reassurance but scaffolding. The CTO explicitly defined a "safe-to-fail" domain, where the team owned internal service logic and data modeling outright, while only public API changes or infrastructure costs above a clear threshold required escalation. Once the boundary was named, the team stopped asking, "Should we ... ?" and started asking, "How fast can we ... ?" The drag of the surprise veto disappeared.

In another organization, a product-led frontend team struggled with constant context drain. They efficiently built features but repeatedly paused to confirm user experience edge cases, worried about violating some unspoken company standard. The fix was deceptively simple: a clear distinction between what was considered standard execution and what counted as innovation. Most of their work fell squarely into the standard category, where the team was fully empowered to decide. With that clarity, reviews vanished, confidence rose, and shipping became a daily habit instead of a weekly event.

A third example came from a platform reliability team operating under constant stress. They knew when error budgets were exhausted, but they hesitated to enforce a feature freeze, fearing political backlash. The CTO removed the ambiguity by codifying error budgets as an immutable policy. When the budget hit zero, the system, not the individual, made the decision. The team no longer had to "police" others or ask for permission. The scaffolding was strong enough to carry the weight of the decision.

Across all three cases, the pattern was the same. Once the decision domains were made explicit, things began to move again. Not because the work got easier, but because the mental tax of uncertainty was removed. Good decision-making structures are like scaffolding: strong enough to support risk and also flexible enough to adapt as the team matures.

Strategic Capability

Getting teams to act without you is a big step. Getting them to think without you is an entirely new level. Tactical execution is essential, and it is essential to use strategic thinking as leverage. A self-managing team that only delivers what it's told is still limited. The goal is not just autonomy; it is also stewardship and ownership. It is the ability to hold the business context and make intentional choices that align with long-term outcomes. Strategic capability does not emerge by accident. It is taught, nurtured, and, most of all, modeled.

Start by involving teams earlier in the planning process. Let them see not just what is being prioritized but also why. Share the company goals and constraints, including revenue targets, customer growth, and budget realities. Treat them like business owners, not just executors. It's amazing how quickly teams rise to the level of responsibility they are given when they understand how their work connects to outcomes that matter.

One useful habit is embedding reflective prompts into your team's planning cycles. Before launching a new initiative, ask, "What are we assuming here?" "What risks are we ignoring?" and "Where does this approach trade off speed for stability, or vice versa?" Some CTOs shy away from these kinds of questions, worried they will slow things down. But what they actually do is increase ownership.

When your team starts thinking like a product manager or an executive, and not just as an engineer, you stop being the only person holding the strategic map. That is the shift: from delivering what was asked to building what is needed.

The Meeting Ecosystem

Meetings get a bad reputation, often deserved. But the problem isn't the existence of meetings—it's their design. In a healthy system, meetings are not a time sink but an alignment mechanism. They help people share context, surface friction, and refocus efforts. However, to do that, they have to evolve along with the maturity of the team.

Self-managing teams need fewer meetings, but those meetings must be better. GitHub's research provides stark evidence: Increasing from 2 to 3 meetings daily causes goal completion rates to plummet from 74% to 14% (Kalliamvakou, 2021). You will know your team is maturing when they start asking, "Do we need to meet for this?" rather than, "Can we meet?" The shift begins by auditing your calendar with a single question: What purpose does this meeting serve? Is it about status? Decisions? Relationships? And who truly needs to be there?

Some CTOs opt for a rhythm of light, asynchronous updates, such as short Loom videos, Slack updates, or shared Notion documents, each paired with more deliberate and focused strategy sessions every few weeks. Others reduce standing meetings in favor of opt-in deep dives, where those closest to the problem invite collaboration rather than passively reporting up.

A CTO coachee eliminated five recurring meetings after simply asking their team to define what decisions were actually being made in those rooms. The answer? None. Just updates that no one remembered a day later. Better meetings mean clearer minds. And when your meetings are designed to support autonomy rather than compensate for its absence, everyone's time becomes more valuable.

Metrics That Matter

It is important to remember that autonomy does not mean abdication. When your team begins managing themselves, you still need to understand what is happening. But the way you measure must shift. In a more traditional organization, metrics often center around throughput: how many story points, how many commits, and how many incidents closed. But those indicators only tell part of the story, and they often incentivize the wrong behavior.

A self-managing team should be evaluated on *capability* and velocity. You want to understand how fast they are moving and how effectively they are operating within a broader system. Useful metrics include the following:

- **Lead time to production:** Not just features, but for learning. Ask, "How fast can we experiment?"
- **Recovery time after failure:** Ask, "How resilient is our system, and how empowered are people to fix things quickly?"
- **Cross-team dependency resolution:** Measure how often you get blocked and how long it takes to unblock yourselves.
- **Decision latency:** Consider how long the team is taking between recognizing a problem and acting on it.
- **Psychological safety indicators:** Question if people feel they can surface the tension, disagree constructively, and take ownership without fear.

These are not abstract ideas. They are indicators of organizational health. When these metrics improve, outcomes tend to flow: There are fewer missed deadlines, less burnout, better quality, and more trust. You do not need to watch everything, but you need to listen to the system.

Refactor in Action

Alex walked into the planning room, laptop under his arm, ready to steer. Except the session had already started.

His directors were huddled around the whiteboard, sketching trade-offs. One outlined the risks of adding a new service. Another pushed back, arguing for simplicity. They'd drawn timelines, cost ranges, and even a draft decision matrix.

Alex slid into a chair quietly, half curious, half uneasy. Normally, everyone waited for him before moving forward. Now, they were doing it without him.

The plan they landed on wasn't what Alex would have chosen. It had rough edges, compromises he would have challenged. But it was coherent. Aligned. And no one had looked to him for the final word.

When the meeting ended, Alex lingered in the doorway. Should he correct them? Step back in? Instead, he said quietly, "Sounds good," and left.

Walking out, he felt a knot in his stomach—equal parts pride and loss. They didn't need him in the way they once had. But maybe that was the point.

Your Way Forward

The highest output of a CTO is not what they build. It is *who* they build. Systems may scale, but people carry the spirit of those systems forward. The moment your team can operate with confidence, clarity, and autonomy in your absence is not the moment you have become unnecessary: It is the moment your leadership has taken root.

You have designed a structure, clarified decisions, and measured what matters. You have made yourself optional because you have deeply invested in capability. That is the shift from bottleneck to builder of builders. From centralized direction to distributed leadership.

However, even the most autonomous teams, if unsupported by a durable culture, will eventually falter. Systems erode. Norms drift. Without reinforcement, even high-functioning teams can begin to lose their edge. This is where the next stage of your evolution begins.

If this chapter was about operationally removing yourself, the next is about ensuring your presence continues symbolically and structurally. It's about scaling your values not by enforcing them but by embedding them into how your team thinks, acts, and chooses every day.

Culture is not what is written on the wall. It is what gets done when no one is watching. And if you want your organization to keep moving in alignment without your constant guidance, then you need to treat culture like code: something you actively design, maintain, and refactor as you grow.

In the next chapter, you will explore how CTOs shape their organization's cultural operating systems (OSs). You will learn how to codify values into rituals, turn documentation into a form of leadership, and recognize the symbolic actions that define what your team believes is rewarded or silently punished.

Building a self-managing team is a powerful achievement. But building one that protects its principles, scales its integrity, and outlives your direct influence? That is legacy.

CHAPTER 9

Culture as Code: Scaling Your Values Without Your Presence

Culture is not what you write on a wall: It is what your team does when you are not there to remind them.

Early in a company's life, culture feels effortless. You do not need to write it down because everyone breathes it. People absorb norms by sitting next to each other, watching how decisions are made, and observing how leaders react to stress. Values and norms are conveyed through proximity, not process.

The founding team shares space, late nights, wins, and war stories. Everyone knows why the company exists, what matters, and what lines not to cross, even if no one has ever written it down. You watch how the founder reacts when a customer churns. You learn how decisions get made by sitting in on three back-to-back meetings. You know what "done" means because you are in the Slack thread when it is debated live. Culture, at this stage, is a living organism shared through osmosis.

Just as importantly, the people who join you early are *drawn* to the rhythm. The pace, the ambiguity, and the energy resonate with them. They join because of what they see, and those who do not vibe with it self-select out. There is no need for a handbook when everyone is swimming in the same current. It works until it doesn't.

The Cultural Operating System

Growth breaks that luxury. As the number of employees increases, new layers appear: more roles, more teams, more managers, and more complexity. You start hiring people who have never met the founders. They were not there when the mission was scrappy and raw, and they are joining something more structured, more stable. Once your team is large enough and complexity increases, culture stops being something people inherit naturally.

This is the inflection point where culture stops being ambient and starts requiring architecture. If you don't design it, you drift. That drift is invisible until it's painful:

- when a team is optimized for speed at the expense of trust
- when a high performer behaves like a jerk, and no one is sure who should say something
- when you realize the values on the wall are not the behaviors in the hall

To regain alignment, you need to think of this as a *cultural OS*, the invisible infrastructure that holds everything together. This will help you make it intentional. Like an OS, it is not flashy. You do not notice when it is working. But everything runs on top of it. Without it, every new app, process, or hire starts behaving inconsistently, and performance degrades. Designing this OS is how you scale culture from something shared intuitively to something reinforced intentionally.

A cultural OS has two parts: *values* (what matters) and *principles* (how that shows up in practice). Too many companies stop at values. They declare, "We value collaboration," but never define what collaboration means in Engineering or Product Development. Does it mean pair programming? Cross-team code reviews? Slower decision cycles to gain broader input?

One CTO I know addresses this by running an annual "values translation" workshop. Each Engineering team takes the company's values and makes them into specific behaviors for their context. The process is rarely smooth at first. Developers don't always agree on what those principles look like in practice, and the discussions can surface real tension about trade-offs, priorities, and past decisions.

This is where the CTO must shift fully into a coaching role. Rather than enforcing answers, they guide the team through disagreement by asking clarifying questions, naming assumptions, and helping the group arrive at shared definitions. By the end of the workshop, the outcome is not just a document. You will have alignment. People can act with confidence, even when no one is watching, because they collectively arrived at shared definitions and now understand what "good" looks like.

Documentation as Leadership

Culture spreads through stories, but it survives through documentation. A growing team is like a growing codebase. Without shared memory, you accumulate "cultural debt." New hires learn hard lessons. Processes get reinvented, and decisions start contradicting each other because no one remembers the reasoning behind them. Good documentation is not a collection of rules. It is an artifact of leadership, designed to capture *why* the team works the way it does, not just *how*.

Documentation is the antidote, but not in the bureaucratic sense. Good documentation is not a list of rules to obey; it's an act of leadership and a thinking artifact. When done well, it preserves what the team does and the principles and trade-offs that shape how and why it operates. It makes implicit wisdom explicit and gives new team members a running start without diluting what makes the team great.

But the power of documentation is not just in what it shares; it is in what it shapes. The discipline of writing is an art form. When a CTO or a senior leader sits down to document how decisions are made, what good looks like, or why something is changing, it forces clarity and vague ideas to sharpen. Reactions become reasoning, and you step out of the chaos and into a posture of reflection.

Without this discipline, you rely on ephemeral memory like a mental caching system. The problem is that while this type of memory is fast and convenient, it is also unreliable. The important context you hold in your head gets overwritten by the next urgent fire, and, just like in computing, overrelying on cache leads to inconsistency, fragility, and eventual failure. Writing, by contrast, becomes your persistent storage: something others can read, improve, and build upon.

One engineering organization used a Claude skill to create a "Principles Playbook" as a companion to its cultural OS. Each principle came with examples: a story about an experiment that failed but taught them something critical, or a decision to delay a launch to run a design sprint because it clarified assumptions. These examples, which served as *lessons learned cards,* gave new hires a living sense of the culture they were joining.

The best documentation evolves. Teams revisit it quarterly, cutting out what no longer fits and adding new lessons. Over time, it becomes less of a manual and more of a mirror. It is no longer a work manual; it is a reflection of how the team thinks, decides, and grows.

Technical Decision Records

If documentation is the story of how your culture behaves, *technical decision records* are the story of how your culture thinks. A technical decision record answers four questions:

- What problem are we solving?
- What options did we consider?
- Why did we choose this path?
- What trade-offs are we accepting?

By writing these down, teams create more than technical memory; they create cultural memory, and they provide much-needed context for AI documentation agents. Future engineers won't just see what was decided. They will also be able to see the values embedded in those decisions. Was scalability prioritized over speed? Was developer experience traded for faster onboarding? These choices reveal what the team believes matters the most.

Werner Vogels at Amazon exemplifies this approach. He proved that for a CTO, writing is the ultimate leverage. By replacing manual oversight with automated mechanisms and documented principles like "Working Backwards," he ensured that Amazon's culture could scale to millions of servers and thousands of teams without losing its soul—or its speed. Similarly, Adrian Cockcroft's journey from Netflix to AWS shows how documented "cloud native" architecture principles can influence millions of developers globally. When leaders write down the why, they aren't just filing a record; they are exporting their culture.

One CTO told me their technical decision library became an unexpected onboarding tool. New engineers did not just learn the architecture: The playbook also allowed them to onboard the organization's mindset. "They started making decisions like we do," he said, "without needing to ask us what to do." That is the power of this tool: preventing culture from becoming folklore.

Ritual Design

If the cultural OS is the foundation, and documentation is the memory, then rituals are the heartbeat. These will be the repeated actions that encode values without requiring explanation. They are how you make culture visible.

One Engineering team runs weekly demos, but they don't just showcase features. Each presenter explains which principle their work embodied. A new API might be tied to "Developer Empathy," while a refactor might connect to "Sustainability over Shortcuts." Over time, these connections become second nature. Values are not abstract. They are lived.

Rituals can also be reflective. Retrospectives are powerful cultural anchors when framed around more than just process improvement. Asking, "What did we learn about how we work together?" brings to the surface patterns that reinforce or challenge the culture you are building.

Designing these rituals starts with identifying what you want your team to remember, even if you never say it again. Then you create practices that make those memories unavoidable. Try this exercise: List the moments in your team's week, month, and quarter where you could anchor value. A daily stand-up might symbolize focus. A quarterly strategy session might symbolize ambition. Over time, these rituals become scaffolding where habits, not slogans, carry the culture.

One of the simplest and most powerful rituals is the art of documenting together *live*. There is something quietly transformative about the moment a team member shares their screen, opens up a blank document, and starts capturing the room's learning in real time. Instead of passively absorbing insights, the team participates in shaping them. The act of writing becomes a collaborative reflection where language gets debated, assumptions get surfaced, and clarity emerges.

These moments do more than produce meeting notes; they build a shared authorship of the culture. When a team collectively refines how to phrase a decision, or how to name a pattern they have seen before, they are not just creating documentation; they are also encoding their shared mental models. They are rehearsing alignment in public and, since this is happening in real time, with everyone watching, the social reinforcement is strong. What gets written gets remembered, and what gets cowritten gets owned.

In a fast-moving company, rituals like these provide cultural gravity. They keep values from floating away and remind the team how the company thinks. When done consistently, they can ensure the cultural scales by rhythm, not by force.

REFLECTION PROMPT

When you are not in the room, what values do your systems, rituals, and decisions express on your behalf?

The Symbolic Actions

Not everything that shapes culture is planned. Often, the most powerful signals are symbolic actions: the things you celebrate, tolerate, and eliminate:

- **Celebration** is obvious but underused. When you recognize someone, tie it explicitly to a value. "Great job shipping that feature" is fine. "Great job shipping that feature while protecting customer trust by catching a subtle edge case" reinforces what matters.
- **Tolerance** is trickier. Every time you let something slide, you teach the team that it is acceptable. A manager who consistently interrupts others. A team that meets deadlines by burning out its members. These moments are cultural crossroads. If you do not address them, they become norms.
- **Elimination** is the hardest and also the most defining. Sometimes, it means parting ways with someone who performs well but poisons the culture. Other times, it is killing a process that looks efficient but undermines trust. These decisions hurt in the moment but pay off in clarity.

Symbolic actions are where belief becomes visible.

Refactor in Action

"Wait—when we say speed, do we mean ship fast… or ship safe?"

The new engineer's question hung in the onboarding room. A few veterans glanced at each other. No one had an aligned answer.

For years, the culture had been instinctive. You learned by osmosis—late nights, war stories, watching how leaders reacted under stress. But now, with 80 engineers and layers of new hires, the rhythm was fading.

Zara felt it too. Values were on the website but not in the work. So, she pulled her teams into a workshop. Instead of slogans, they translated values into behaviors. Speed meant feature flags, not hacks. Trust meant open incident logs. Ownership meant writing decisions down. By the end, they had a messy, living playbook—something concrete to point to when instincts blurred.

Two weeks later, pressure hit. A critical launch was slipping, and Product pushed to skip a security check. The team didn't escalate to Zara. They opened the playbook, logged "ADR-027: Delay one week to protect trust," and owned the decision.

At the next all-hands, Zara didn't celebrate shipping speed. She celebrated that choice. The story spread faster than any memo.

That's when she knew: She wasn't carrying the culture anymore. The culture was carrying them.

Your Culture, Written in Code

Building self-managing teams is a powerful milestone. But autonomy without alignment is fragile. What holds it together is culture, not as slogans but as systems. As code. Your job is not to memorize and recite the values for everyone else. Your job is to embed them in how your teams work, think, and remember. Culture becomes durable when it is systematized, ritualized, and protected.

As you design your cultural OS, ask yourself these questions:

- *If I disappeared for six months, would this culture survive?*
- *Which rituals reinforce our values, and which silently erode them?*

- *What is one symbolic action I could take this month to make our priorities undeniable?*

Because the real test of culture is not whether people can recite words. It is whether, when faced with a hard choice, they instinctively make the decision that feels "obvious," because it reflects who they are as a team.

Building self-managing teams allows your organization to run without you. Designing a cultural OS ensures it runs true to its values. But culture is not static; it has to adapt, and the forces shaping it are accelerating faster than ever. Few forces are reshaping the CTO's world as profoundly as AI.

AI is not just a new tool on the stack; it's also a new kind of leverage. It challenges how we think about productivity, creativity, and even leadership itself. Handled well, it can multiply everything you have built, making your systems faster, your documentation richer, and your teams more capable. Handled poorly, it can amplify chaos, distort values, and erode trust.

The next chapter explores how to approach AI as a force to help you lead with clarity. We will look at what it means to be an *Augmented CTO*, where you will use the tool while keeping human judgment, ethical guardrails, and cultural backbone that make your organization distinctly yours. AI, when applied wisely, is what will give that identity its greatest reach.

CHAPTER 10

The AI Advantage: The Ultimate Multiplier

AI is not here to replace you; it will multiply you if you lead instead of chasing it.

Throughout this book, you have explored different forms of leverage: delegation, systems, and culture. Each has expanded your impact by reducing the dependency on your direct effort. However, AI represents a fundamentally different kind of leverage. It allows you to reduce effort and change what it even means.

In the past, scale came from adding people or building better processes. AI allows scale by compressing time and amplifying cognition. Tasks that once took hours will take minutes, and insights that required specialized expertise will become accessible to anyone with the right prompt. For the CTO, this is not just efficiency; this is a fundamental shift in what's achievable. Ask yourself, *What could my team achieve if their mental bandwidth was not spent on the repetitive, the routine, or the barely manageable complexity?*

Research from Gartner indicates that organizations with mature AI practices are realizing an average of 15.8% revenue increases and 15.2% cost savings (*Gartner Predicts 30%*, 2024). However, the gap between "experimentation" and "value" is widening: Gartner also predicted that 30% of generative AI projects would be abandoned by 2025 if they weren't tied to specific organizational leverage. At the developer level, the impact is visceral: GitHub Copilot users finish tasks 55% faster, with 75% reporting they can focus on more satisfying work (Kalliamvakou, 2022). This isn't just efficiency; it is the reclamation of cognitive capacity

At the same time, leverage will magnify everything, including dysfunction. A team without clarity will only make bad decisions faster. An organization without strong values will automate away its integrity. Remember that AI will not fix what is broken, but it will scale what is already there. Therefore, the first thing you should ask is not *which* tool you should use. Change that for: "*How* do I want to multiply?"

The Augmented CTO

The most effective leaders I have seen do not treat AI as a shiny object or a threat. They treat it as a helper that will expand their capacity for judgment, creativity, and connection. For CTOs, that begins with writing. Not because AI needs it, but because *you* do.

Tools like large language models are only as good as the inputs you feed them, and clarity of thought is the ultimate input. If you already keep detailed documentation, AI can become your amplifier, summarizing complex discussions, generating first drafts of proposals, or highlighting patterns across technical decision records.

REFLECTION PROMPT

If AI scaled your decisions today, would you be proud of what it amplified? What conversation do you need to start this week to make sure speed doesn't outpace trust?

One of the most powerful tools I have introduced to CTOs is what I call the *CTO Notebook*, which you can find online at ctonotebook.com. At its simplest, it's a living repository where you, as a technology leader, document not only what you decide but also why. Architectural principles, leadership reflections, decision-making trade-offs, and even personal learnings all belong here.

At first, the CTO Notebook often starts as a private archive, capturing reasoning that would otherwise get lost in the rush of scale. But the real power comes when it evolves into an augmented system. One leader I worked with connected his notebook to an AI assistant, transforming it from a static document into a dynamic decision partner.

Suddenly, he could query past decisions to recall the context between them, generate onboarding guides for new engineers based on his own principles, simulate strategic scenarios using his historical thinking as input, and provide his team with clarity and continuity without having to be in every conversation.

This is what it means to operate as an Augmented CTO. It does not mean that you will outsource your thinking; rather, you are structuring it so that AI can extend what you already have. The CTO Notebook becomes an operating system for your leadership, one that compounds value over time. The discipline of writing, of codifying what you think and why you think it, is leverage, and the augmentation is the multiplier. Together, they turn ephemeral memory into permanent strategy.

But here is the key: Your own augmentation as a leader is only the starting point. The real test of an Augmented CTO is whether your organization is equally ready to harness AI. Remember that even if you build clarity, context, and systems for yourself but your company pours AI into confusion and chaos, the leverage disappears.

The Organizational AI Readiness Assessment

Every company wants to "do" AI, even when they are not ready. Implementing AI without preparation is like pouring water into broken pipes; it will get everything wet despite spreading faster. A recent McKinsey study found that nearly 80% of companies using AI "report no significant bottom-line impact," and the broader failure rate for AI initiatives sits between 70% and 85%, significantly higher than standard IT projects (*Enterprise Generative AI 2026*, 2025).

Gartner predicted that by 2025, 30% of generative AI projects would be abandoned entirely after the proof-of-concept phase (*Gartner Predicts 30%*, 2024). The difference between the 15% who succeed and the 85% who fail is rarely the technology itself but the organizational readiness to wield it. Spending was nonetheless expected to approach $62 billion in 2025, even as 42% of corporations abandoned AI pilot projects in 2024 (Tomase, 2025).

Just as you had to create the scaffolding for your own thinking with the CTO Notebook, your company needs the same to handle AI responsibly and effectively. For this reason, it is essential to make an assessment and understand if your company is ready to start using it. A practical readiness assessment asks three questions:

1. Do you have clean, accessible, and relevant data to support meaningful AI use cases?
2. Are your workflows documented enough that AI agents can enhance them, or are they so ad hoc that automation would create more chaos?
3. Does your team trust new tools? Do they understand AI's limitations as well as its potential? Or will fear and skepticism erode adoption?

A midstage SaaS company I worked with learned this the hard way. It invested heavily in AI-driven customer support before cleaning up its ticketing data. The model "learned" to reproduce the inconsistent responses their agents had been giving. The result was *faster bad support.* They had to pause, fix the foundation, and then reintroduce AI with more intentionality.

Keep in mind that you should start small, or maybe even wait until the organization is ready. Let your team experiment on less-critical systems before implementing them in core areas.

Strategic AI Integration

AI adoption often swings between those who believe AI will solve everything and those who believe that it will ruin everything. The truth lies in developing a strategy for using the technology to multiply your strengths, and ignoring it when it distracts you from them. A quick framework you can use is to map them by impact and confidence:

	Low impact	High impact
Low confidence	Ignore for now	Experiment carefully (using predictive models for customer churn, for example)
High confidence	Use opportunistically but do not overinvest (ex: AI to draft internal meeting notes)	Automate or augment aggressively (such as AI-assisted testing to catch known bug patterns)

Strategic integration also means thinking beyond productivity, where AI can shape how the organization learns. Imagine AI tools that surface patterns across retrospectives, identify latent dependencies in your architecture, or help you simulate the cultural impact of policy changes. Keep in mind that the idea is not to have an AI strategy but rather to use AI to accelerate, not replace, the system you are building.

The Ethical Imperative

Power without principles is fragility in disguise. AI, by its very nature and power in the form of scale, speed, and influence, allows decisions to be made at a velocity and scope that no human could match. However, those decisions, when shaped by flawed data or unchecked automation, can do harm faster than anyone can respond.

A CTO I know deployed an AI-driven anomaly detection system to flag potential breaches in real time. On paper, it was flawless: The system reduced alert fatigue for engineers by automatically closing "low priority" incidents. But buried in those dismissals was a subtle but sophisticated attack pattern that slipped through undetected. The AI had learned to ignore signals that were historically noise and, in doing so, created a blind spot that hackers exploited. What looked like automation efficiency turned into an expensive vulnerability.

The ethical imperatives for CTOs are more operational than philosophical. It lives in the models you choose, the data you train them on, and the safeguards you build around their outputs. Trust remains a critical barrier to adoption, particularly among those closest to the code. According to Stack Overflow's 2024 survey, only 43% of developers trust the accuracy of AI tools, a sentiment that is declining rather than improving ("Developers Want More," 2025). Furthermore, 79% cite misinformation as their top ethical concern. If your engineers—the very people building the systems—do not trust the output, you cannot expect the broader organization to follow.

If your AI agent recommends hiring a candidate, a marketing strategy, or even a security decision, you should be ready to hold those who are accountable. What assumptions is the system making, and, most importantly, are you comfortable owning them?

Start with questions that go beyond compliance checklists:

- Are we transparent about where and how we are using AI? For example, if your customers or team do not know AI is involved, what does that say about your design?
- Do we have active monitoring for bias, misuse, or unintended consequences?
- Are we prepared to shut down an AI feature, despite how profitable it is, if it violates the organization's principles?

Consider the cautionary tales:

- A widely publicized recruitment algorithm began excluding women from technical job recommendations because it was trained on historical data that reflected a biased past (BBC News, 2018).
- A content recommendation engine, optimized purely for engagement, started surfacing increasingly divisive material, damaging both public trust and the platform's reputation (Nikos-Rose, 2023).

In both cases, the system "worked" but did so without ethics, making it destructive.

This is why AI decisions must be directly connected to your cultural OS. If your values say "people first," that principle must be encoded into how your models weigh accuracy versus harm and efficiency versus fairness. When you set those boundaries up-front, you do not have to question every trade-off in the moment; the system will be making better decisions, because you have already taught it how to do so.

Yes, it may cost speed, and it may mean saying "no" to features others rush to release. However, it also protects something more valuable: the trust your customers, teams, and peers place in your leadership. Ethical leadership is slow when it needs to be and bold when it must. In an era of AI acceleration, the CTO who balances both becomes a responsible technologist.

The Human Element

One of the many things that will always define the human role in leadership is responsibility. Machines can generate options, simulate outcomes, and even optimize for patterns we would never see ourselves. But they do not own the consequences of those choices. They do not sit across from a disappointed customer, look a struggling engineer in the eye, or stand before a board to explain a bet that went wrong.

What endures and cannot be outsourced is the human ability to interpret, to weigh trade-offs not only in terms of efficiency, but also in terms of trust, values, and long-term impact. The algorithms may accelerate decision-making, but the meaning of those decisions, and the accountability they carry, still belong to you.

As agentic AI takes on more tasks, your role as the CTO will change. You become less of a mechanic and more of a meaning-maker. You interpret, prioritize, humanize, and help your team see what is possible and also what is right. Your role will become even more strategic than before.

This is where self-managing teams and strong values stop being "soft" practices and start becoming survival infrastructure. In a world where decisions can be made in a fraction of a second, culture will be the organization's guardrail. It ensures your teams, when empowered with AI, will still move with purpose and clarity, ethically, and using their human features to make important decisions.

Paradoxically, the more we adopt AI, the more essential our humanity becomes. I mean this not in the ethical sense but in the practical one. The best teams will be those that foster *human* qualities such as

- **curiosity** to question the outputs, not just accept them.
- **empathy** to understand how those outputs affect real people.
- **judgment** to know when automation should yield to conversation.

I have seen CTOs light up when they realize this shift. AI does not have to erase your identity if you know how to use it to elevate it. When the mundane is handled by machines, what remains is the distinctly human work: mentoring leaders, refining vision, building trust, and stewarding a culture that endures change. AI will not replace you if you accept its invitation to lead at a deeper and more consequential level.

Refactor in Action

The first demo floored everyone.

Maya sat at the head of the table as the AI-generated road map flicked across the screen. Release notes polished, customer feedback neatly summarized, sprint plans mapped three months out. The exec team leaned forward, murmuring approval.

"This is incredible," the CEO said. "Weeks of work, done in minutes."

Maya smiled, but her chest was tight. She had been up half the night feeding backlog data into the model. It felt like magic—until the emails started pouring in.

One customer snapped, "Why are you prioritizing features we begged you to retire?" Another pointed out that the new integrations would put them out of compliance. Meanwhile, engineers grumbled that the AI's polished summaries made them look careless—like their own voices had been erased.

By Friday, Maya's inbox was on fire. The road map had scaled beautifully—just not the right things. The AI hadn't gone rogue. It had simply learned from what was already there: messy data, half-documented decisions, and unspoken values.

And this was the harder truth Maya hadn't really accounted for: Aligning an organization is already a mammoth effort, even without AI in the mix. Opinions on how AI *should* be used were deeply divided between speed and safety, augmentation and replacement, trust and control. Navigating those tensions was a leadership problem.

She stared at the dashboard late that night, the cursor blinking at the top of a fresh prompt. For the first time, she wasn't sure what to ask.

Because the question wasn't, "What can AI do for us?" It was, "What exactly are we asking it to multiply?"

Your Final Leverage Question

Remember that the essence of the CTO's role has not changed and remains understanding objectives, stewarding the budget, and building the technology. However, keep in mind that the nature of the technology is evolving, and today it learns, adapts, suggests, and decides, meaning that your leadership must evolve with it. You will need to transform from operator to architect, from technologist to ethicist.

AI is your most powerful multiplier, but it will depend on *you* as to what it will multiply:

- Will it scale your best thinking or your unconscious habits?
- Will it reflect your deepest values or your fastest instincts?
- Will it help your team become more human or just more efficient?

The most important lesson here is that the choice is not the tool you will use but *how* you lead its use. AI can extend your reach and reclaim hours from repetitive work, but it cannot manage your attention for you. It can free your calendar, but only you can choose where that freedom flows.

This is the paradox: AI gives back time, but unless you design it with intention, that reclaimed time gets consumed just as quickly as before, by context switching, shallow work, or reactive availability. For this reason, in the next chapter, we will shift from algorithms to something more elemental: your time as a design principle. You will learn how to reclaim time for sustainable leadership, because no matter how you leverage what you create, your most limited asset remains the same: *attention, energy,* and *focus.* Read on to learn how to protect them.

PART III

TIME-HACKING FOR CTOS

CHAPTER 11

Reclaiming Time: Your Scarcest Resource

Your hours are finite, but your focus is fractal. Mastering attention is your true superpower.

CTOs often sit in the strange intersection between surplus and starvation. You might have back-to-back meetings, DMs piling up, and inboxes flashing with urgency, but none of this seems to make sense, as your most strategic initiatives remain untouched. You are busy, but not always effective: busy, but not always present. There are hours in the day, but you have lost time.

This is the time paradox: The more senior your role, the more expansive your impact. At the same time, you feel that you have less control over your own schedule. One CTO I coached put it this way: "It is like being in 100 rooms at once, yet unable to move anything forward in any of them." That is not a calendar issue; it is a focus failure, one of the CTO's most relevant efficiency-killers.

Therefore, let's start with this truth: You are likely not under-resourced; you are overcommitted. Let's fix that.

The Attention Economy

As CTO, your attention is among the most expensive and high-leverage assets in the organization. Every context switch, every interruption, and every meeting you attend without full presence comes at a cost, not just in minutes but also in strategic clarity, mental sharpness, and leadership momentum. This isn't about being busy. It's about being *fragmented,* keeping in mind that, in a scaling company, fragmentation compounds.

Attention is under constant assault from interruptions and context switching. Dr. Gloria Mark's decade-long research at UC Irvine demonstrates that frequent context shifts create attention residue: the cognitive leftover from a previous task that lingers as you attempt to engage with a new one (Mark et al., 2008). Her studies show that knowledge workers rarely achieve true multi-tasking. Instead, the cycle rapidly switches between incomplete thoughts, with each switch reducing focus and depth of thinking. The cumulative effect is a steady erosion of cognitive capacity, making every interruption a hidden cost to productivity and strategic performance.

Your job, then, isn't merely to manage your time; it's to *protect your focus.* Because time can be filled. Calendars can be optimized. But focus? That's your multiplier. And when it's diluted, the systems you're supposed to be scaling start to wobble. It's not about guarding your calendar like a fortress. It's about designing your week like a high-availability system: resilient, intentional, and optimized for throughput, not just uptime.

However, there is a trap: Because you are capable, you will get pulled into everything. Escalations, architectural debates, recruiting, one-on-ones, incident reviews, and even product brainstorming are a part of your routine. At some level, you enjoy it, although this does not mean that just because you *can*, you *should*. Attention is not evenly distributed, as some moments require the deepest clarity and others do not require you at all. Learning to tell the difference and designing for it is what separates reactive leaders from strategic ones.

Context Switching: The Hidden Tax

When you switch contexts, time is not the only thing you are losing. You must also consider the depth that is being left out. Research shows that task switching can reduce productivity by up to 40% and, more importantly, it fragments your attention in a way that makes you forget not only what you were doing but why you were doing it: "Although that shouldn't surprise anyone who has talked on the phone while checking E-mail [sic] or talked on a cell phone while driving, the extent of the problem might come as a shock" (APA, 2006).

As CTO, this becomes exponential, because you are not just switching between projects but also changing modes. You need to constantly change between visionary mode, mentorship mode, problem-solving mode, and decision mode, among others. Each of these modes has a unique rhythm, energy profile, and mental setup, and changing between them requires a corresponding change:

- **Visionary thinking** requires expansiveness and abstraction.
- **Mentorship** requires emotional bandwidth.
- **Problem-solving** requires crisp focus and domain depth.
- **Decision-making** requires decisiveness and information synthesis.

Rapidly cycling between them does not just cost minutes; it also creates cognitive dissonance. It prevents you from fully arriving.

One CTO I worked with did a brutal exercise: She color-coded her calendar by cognitive mode and was shocked by the result. She was switching modes every 30 to 45 minutes for days on end. Her week looked like a patchwork quilt of tasks and transitions, with no depth blocks longer than an hour. She was tired and fragmented.

The insight? Your brain is a sequencer, not a multiplexer, and when you try to parallelize human cognition, everything slows. The idea that "your brain is a sequencer, not a multiplexer" is a metaphor often used in neuroscience and cognitive science to describe how the brain processes information. It implies that the brain primarily functions by sequentially processing information rather than by simultaneously handling multiple streams of information, like a multiplexer. You feel like you are spinning plates, and that they are all about to crash.

The Deep Work Deficit

Although many CTOs I meet believe they lack time, this is untrue; they are, in fact, lacking depth. Shallow work is easy to fall into. It gives us the dopamine of responsiveness: Someone Slacks; you reply. A text message pings; you respond. A stakeholder calls; you jump. It feels active, but these activities are rarely strategic. Deep work, by contrast, requires stillness, demands intention, and appears when you

- architect a reorganization that enables future scale.
- write a memo clarifying your long-term technical bet.

- develop a decision framework that your team can use independently.
- reflect on where you are unconsciously becoming a bottleneck.

The list you have seen is not made of items that you can cram between meetings. They require time *before* the time: the immersion, the emotional capacity, and all the characteristics that brought you to where you are. The more complex your organization, the rarer this becomes unless you build for it.

Our CEOs place a high premium on availability and past experiences, which is why we CTOs must be available 24/7. This requires coaching our peers that not being available does not reflect on our commitment. We need to be intentional. But here is the challenge: Most CEOs, boards, and even peers have been conditioned to see *availability as a proxy for commitment.*

You must consider that deep work is a core leadership contribution, not a luxury, and when it's missing, you start to mistake volume for value. However, you cannot out-meet your strategic debt, as only depth will be able to close this gap.

The Time Audit

If you are serious about reclaiming your focus, you should start with data. A time audit is a visibility tool, not a guilt trip, as you cannot change what you can't see. Here is a simple process:

Step 1: For 5 business days, log your time in 30-minute increments. No rounding. Just note what you did.

Step 2: For each block, rate:

- *Energy (1–5):* How energized or how drained did this activity make you feel?

- *Leverage (1–5):* How impactful was this toward your organization's long-term goals?

Step 3: At week's end, ask yourself these questions:

- *Where is my time going by category?*
- *What patterns am I repeating unconsciously?*
- *How much of my week is low-leverage, low-energy?*
- *Where am I pretending to be strategic but actually being reactive?*

One CTO found that their calendar and inbox had become an artifact of everyone else's priorities. Only 14% of the time was truly high leverage. The rest? Meetings that did not need him, Slack rabbit holes, approvals he did not trust his team to own. This was his wake-up call, and yours may look different. But the signal is always the same: You are not where your greatest impact lives.

REFLECTION PROMPT

What's one recurring activity that drains your energy and delivers little strategic value? What would happen if you eliminated or delegated it this week? When was the last time you spent 90 uninterrupted minutes thinking deeply about your organization's future? Schedule that time, now!

The Reclamation

Reclaiming your time does not mean isolating yourself. It means recentering your leadership around the things only you can do.

After an audit, here are a few actions you can take to optimize your time:

- **Eliminate meetings that do not create leverage.** Ask yourself, *Does this grow a leader, accelerate a decision, or clarify a strategy?* If not, cancel it.
- **Design deep blocks.** Two 90-minute blocks per week, minimum. Protected. No meetings or context switches.
- **Batch the noise.** Slack and email windows twice a day, and allow async work to be async.
- **Guard your mornings.** Do not wake up in the firehose. Start with clarity, including practices such as journaling, planning, or your highest-leverage task.
- **Name your mode.** Start your day by naming what kind of leader you need to be that day. It frames your choices, your meetings, and your posture.
- **Say no.** Generously, strategically, and with context. "No" is the boundary that creates your leadership edge.

This is design work, not hustle, with exponential benefits. As your attention compounds, so does your impact.

Refactor in Action

Carlos sat in the boardroom, eyes fixed on the slides. On the surface, he looked engaged. But in his mind, notifications from an outage kept buzzing. He had half-written a message to his ops lead, replayed three different debugging scenarios, and silently cursed himself for not muting his phone.

When the CEO turned and asked, "Carlos, do you see Engineering supporting this new Product push?" he froze. He had heard the words but not the meaning. He had no context. His silence lasted just long enough to make the room uncomfortable.

Later, in the hallway, the CFO joked, "You looked like you were buffering." Carlos forced a smile, but inside he knew the truth: He hadn't been present.

He had been in the room. He had cleared his schedule. But his attention was still shredded across half a dozen invisible fires. His calendar wasn't the problem. His attention was.

That night, he replayed the moment again, not because of embarrassment but because he realized what was at stake. The company didn't need more of his hours—it needed his undiluted presence in the moments that mattered.

Your Attention Is the Architecture

Time is finite, but how you allocate attention (invest depth, clarity, and presence) becomes the architecture of your impact. The more you reclaim focus, the more leverage you build. Not through effort, but through intention. Not by working longer, but by working aligned.

You have already begun this shift: designing self-managing teams, scaling your cultural OS, and protecting deep work. But now it's time to make it real, not just as a theory but also as a calendar. Because the next evolution of your leadership is not about abstract priorities but about concrete patterns in a week that reflect your strategy and a schedule that protects your role.

In the next chapter, we will build exactly that: *your ideal week.* Not the perfect week for someone else, but the one that makes your leadership sustainable, focused, and high leverage.

CHAPTER 12

The Ideal Week: Engineering Your Perfect Schedule

Build your week like you would build a system: for reliability, clarity, and peak throughput.

For all of us, the week goes away in a blink, and nothing seems to go as planned: You start Monday with a plan, and by Tuesday, that plan has been shredded by developer discord, long-winded exec sync-ups, draining decision-making, and burning up creative energy. By Friday, you are wondering where your strategy went, or worse, if you even did anything that mattered. That feeling is a scheduling architecture problem and not a time management problem.

In the same way that you would not launch code without thinking about system performance, you should not structure your week without optimizing for energy, focus, and throughput. The ideal week will be created if you solve its design problem. As the CTO, you are more than capable of reaching this objective.

The Power Schedule

Before embracing the scheduling architecture mindset, think about how most executives live in a reactive loop. Your calendar is a patchwork of meeting requests, Slack-driven decisions, and placeholder blocks that never hold. You wake up to a full day and go to bed feeling like the day could have been more productive. This is your *exhausting* default operating mode.

However, the shift begins when you see your time as architecture. Not just something you inhabit, but also something you intentionally design. The Power Schedule starts with understanding your energy profile. Are your mornings more focused? Do you hit a creative stride in the afternoon? When do you hit cognitive fatigue? Mapping this honestly allows you to place demanding work where it has the highest chance of success, rather than forcing focus when your system is already depleted.

Equally important is giving the week itself a shape. Setting one or two clear goals at the start of the week creates a directional anchor for everything that follows. These goals are not task lists; they are outcomes. A good weekly goal is specific enough to guide decisions but narrow enough to keep focus. "Finalize the technical approach for X" or "Create alignment with Product on Y" provides a direction you can check as the week unfolds. When a new request arrives, you can ask a simple question: *Does this move the week's goal forward, or does it pull me off course?*

This practice does more than improve productivity; it reduces decision fatigue and restores a sense of control. When you know what success looks like for the week, interruptions lose some of their power, and unfinished work becomes easier to triage without guilt. Over time, weekly goals become a feedback loop: They help you notice when your time architecture is aligned with your priorities and when it quietly isn't.

Design your schedule around energy, not only hours, such as

- **mornings** for high-focus and creative work;
- **middays** for meetings and collaboration; and
- **afternoons** for reviews, reflection, or wrap-ups.

Instead of reacting to everyone else's requests, anchor your week around your most important work. Protect your mornings if that is your peak clarity window, and avoid scheduling meetings during your natural deep work zone. Another idea is to rebuild recurring meetings around your optimal flow.

Think of your schedule like a system with constraints: Your energy and focus are the bottlenecks. A well-designed Power Schedule routes the most critical tasks through your highest-leverage hours. This transformation is not instant, and it takes experimentation, boundary-setting, and a few hard conversations. At the same time, once you reclaim your week's architecture, your impact will compound. You stop leaking energy and start leading from clarity.

Theme Days vs. Mixed Days

One of the most effective ways to reduce context switching is to cluster similar types of work together. When you try to move between building small prototypes, strategy sessions, executive updates, and hiring interviews in a single afternoon, your cognitive system never fully engages. The residue from one mode leaks into the next, preventing you from completely concentrating. When this happens, *theme days* become powerful.

Suppose you have just come off a call with a furious customer, one who insists that your product nearly cost them a major account. You hang up, rattled, adrenaline still surging. Ten minutes later, you walk into a hiring interview with a bright-eyed Engineering candidate. Instead of showing curiosity and warmth, you catch yourself sounding defensive, overly skeptical, and even dismissive. The candidate senses it. You sense it and leave the conversation, wondering why you said what you said. This is not you.

That is the hidden cost of unstructured context switching. Emotional residue lingers. Stress from one domain contaminates your presence in another. Theme days counter this by creating smoother boundaries where strategic meetings are placed together, one-on-ones are clustered, and deep work is reserved for blocks where you don't have to shift into a different emotional register every hour. By aligning similar modes, you reduce the friction of transition and protect your productivity and your presence.

So, I know where you think I'm going with this. You have probably read a dozen books or listened to countless podcasts about blocking time on your calendar into neat little slots: 30 minutes for emails, 60 minutes for a meeting, or another hour to "focus." This is not what I am talking about. What I mean is that instead of designing your week by time slots, you should design it by purpose. Each day carries a theme, so to speak, which will be a mental mode you stay in for the majority of the day.

A CTO like yourself might design the week like this:

- **Monday:** Deep work, strategy documents, and planning
- **Tuesday:** Engineering syncs and team execution
- **Wednesday:** 1:1s, coaching, and mentorship

- **Thursday:** Executive alignment, board prep, and cross-functional updates
- **Friday:** Reflection, writing, and long-term thinking

Of course, specific themes will vary depending on your role, stage, and team size. The point is that by aligning days around purposes, you can create smoother rhythms. You walk into each morning knowing what kind of presence will be required of you, and you protect yourself from the dissonance of switching between "visionary CTO" and "bug-triager-in-chief" in the same afternoon.

Before theme days, your schedule likely looks like a patchwork quilt with its different and scattered colors. You start the day writing a strategy memo, jump into a product fire drill, coach a junior engineer, and then lead a hiring interview. By the end of the day, your brain is mush, and your outputs are shallow.

After implementing theme days, the picture changes. It will bring more rhythm to your work schedule, as you will wake up knowing the mental gear you are shifting into. Your decisions become sharper because you are fully inhabiting the role that day demands. Team members know when to engage and when you are unavailable, bringing higher clarity, deeper impact, and lower stress.

That said, not every leader thrives with strict theme days. Some CTOs perform better with mixed days, where they blend meetings, solo work, and strategic reflection in tighter blocks. If variety energizes you and you can clearly switch, a well-structured mixed schedule can work beautifully.

The key to mastering this is intentionality, since whether you lean toward a theme or a mixed schedule, the trap is defaulting to chaos. Try both approaches for two weeks each and measure your energy, focus, and progress, letting data drive your design. Your week should be a multiplier for impact, not just a container for work, so design accordingly.

Meeting Taxonomy

Meetings are like functions in a codebase: They should be single-purpose, efficient, and limited in scope. When organizations prune this codebase, the results are immediate. Companies like Asana and Shopify have implemented strict no-meeting days, reporting productivity boosts and measurable reductions in employee stress. This isn't just about clearing a calendar; it is about cognitive preservation.

Executives who protect large blocks of uninterrupted time consistently show an improvement in focus and strategic output. However, in most CTO calendars, meetings blur into a blob: status updates pretending to be strategic conversations, check-ins that meander, and high-leverage discussions squeezed between context shifts.

A *meeting taxonomy* is a simple structure that classifies meetings by type. Just like you would refactor a monolith into microservices, breaking down your meeting load by category helps you optimize to increase efficiency and clarity. Start by defining the categories:

- **Direction meetings:** Strategic planning, architecture reviews, and road map design. These require deep preparation, clear objectives, and key decisions.
- **Execution syncs:** Sprint planning, bug triage, and backlog grooming. These can often be shortened, batched, or delegated to tech leads.

- **People meetings:** 1:1s, coaching sessions, and hiring interviews. These demand presence, listening, and emotional bandwidth.
- **Surface scans:** All-hands, demos, and updates. If no decisions are made and no input is required, consider making them asynchronous.

Before applying a taxonomy, you will likely feel that your calendar is somewhat fuzzy, making you unsure where to focus or how much to prepare for each meeting. After taxonomy, each meeting earns its slot, and you will be able to enter with the right mindset and exit with outcomes. Use this taxonomy to delegate execution syncs to empower team leads, protect direction meetings with deep prep time, give people meetings the emotional presence they deserve, and turn surface scans into Loom videos, Slack updates, or shared documents.

Color-coding your calendar by meeting type can create immediate awareness. More red than blue? Too many updates, not enough strategy. Don't just have fewer meetings; have better ones with clear roles and clearer value.

The Sacred Blocks

Sacred blocks are the immovable objects in your calendar, and they must be substantial. You cannot schedule flow, but you must schedule the time required to protect your highest-leverage work. If everything is flexible, everything becomes reactive. Reflection of these changes happens in transit, and strategic thinking is perpetually "next week," and the result is that you are always busy and rarely impactful.

Sacred blocks flip this, since you carve out time *first*, not last. You will define the work only you can do and then create the time and space to do it. Here are some ideas to start creating your blocks:

- **Deep work block (90 to 120 minutes):** Should happen ideally two times per week. No meetings, no pings. Use it for design, writing, and systems thinking.
- **Leadership reflection block (60 minutes):** Once per week. Check in on your team's health, spot brewing issues, and course-correct proactively.
- **Vision alignment block (30 to 60 minutes):** Zoom out. Consider what is next, what is missing, and what is possible.

Communicate clearly to your team: These are not open slots. They are your core work, so make them visible, color-code them, and label them unapologetically. Here is a sample layout:

Day	Morning	Afternoon
Monday	Deep work	1:1s/mentoring
Tuesday	Product strategy	Engineering syncs
Wednesday	Deep work	Hiring and reviews
Thursday	Executive alignment	Writing and docs
Friday	Vision block	Decompression

After implementing sacred blocks, one CTO reported, "I got more done in six hours than I used to in three days." Why? Because high-leverage work is about uninterrupted clarity, not volume, making you a responsible, not indulgent, leader.

The Rule of Thirds

Once you have established a Power Schedule, clarified your meeting taxonomy, and implemented sacred blocks, the question becomes: How do you maintain long-term balance? That is an excellent question, and that is where the rule of thirds comes in. Think of your ideal week as a resource allocation model so that you can decide when your presence creates the most leverage. The rule of thirds helps codify this by dividing it this way:

- **One-third, strategy:** Including long-term planning, architectural vision, technical road mapping, and writing. It is time spent designing the future.
- **One-third, people:** Coaching, 1:1s, hiring, team calibration, and culture building. This is how you amplify your impact.
- **One-third, execution:** Staying close to delivery, reviewing systems, unblocking teams, and ensuring alignment.

Before applying this model, you might find that you are heavily tilted, often spending most of your time in execution or firefighting, with little left for strategic work or people development. That imbalance is not sustainable, since it breeds reactivity, slows down leadership growth, and, ultimately, makes you the bottleneck. After implementing the rule of thirds, your calendar will become a reflection of your actual role, not just your historical habits.

Like any system, the ratio might flex in high-pressure moments. But as a default, it is your operating rhythm, which should be revisited quarterly. Ask yourself, *Did this past month reflect a healthy third–third–third distribution?* If not, why not? Your leadership scales when your time allocations reflect your true leverage.

REFLECTION PROMPT

Where does your current week leak the most energy? Sketch your calendar on paper. Label each block: Strategy, People, or Execution. What's the ratio?

Transition Rituals

You have finally architected your ideal week, but there is still one final catch: Context switching will still happen, even if you have a structured schedule. The difference between a productive switch and a cognitive pile-up? Rituals. Transition rituals are short, intentional actions that help you close one loop and fully enter the next, helping you switch between modes.

Before you jump from a deep writing session into a stakeholder call, you need to tell your brain that it is no longer sorting through paragraphs. If you do not, you will be physically present but mentally elsewhere. After you implement these rituals, you can take five minutes to write a summary of what you just completed, close all the tabs, take a breath, and review your goals for the next meeting. Now, you are arriving with full clarity.

Simple but powerful examples you can use include the following:

- **Reset notes:** Keep a physical notepad next to your keyboard and jot a quick note at the end of a work block: What did I finish? What is next? The tactile act of writing and physically checking items off with a pen provides a small but meaningful sense of closure that helps your brain reset.
- **Physical anchor:** Stand up, refill water, and walk for two minutes. Movement clears residue.

- **Mental shift question:** Ask yourself, *What does this next mode need from me?*

My strategy was to place a sticky note on their monitor with this prompt: *What game am I playing right now?* It reminded me to fully shift from builder to coach, from coach to executive, and from executive to visionary. The more fluid your transitions, the more powerful your presence becomes. Think of it as bringing the right mindset at the right moment to make the right decisions.

Refactor in Action

Jonas's calendar looked like a losing game of Tetris. Blocks of color stacked on top of each other, crammed into every hour, a patchwork of half-finished priorities. From a distance, it looked impressive—a leader in constant demand. From inside, it felt like drowning.

Monday was typical. He opened a strategy doc at 8:00 a.m., got eight minutes in, and then was yanked into an Engineering sync. By 9:15, he was firefighting a customer escalation. At 10:00, he was already nodding through a candidate interview. By noon, he had switched mental gears five times but hadn't created anything of substance.

The crack came during board prep. The slides were pristine, the metrics precise. But when the CEO asked, "What's your conviction on this architectural bet?" Jonas froze. He had surface knowledge but no depth. He'd never given himself the time to think it through. His week was full, but it was hollow.

That night, he stared at the rainbow quilt of his calendar and finally admitted: The system was broken. So, he rebuilt it. Fridays went dark—strategy only. Wednesdays became people days. Mornings before 11:00—untouchable. Two 90-minute blocks boldly labeled: DO NOT BOOK.

The first weeks were messy. Project managers grumbled. A colleague pinged: "Can't we just grab a quick Friday slot?" He held the line. Slowly, a rhythm formed. By the third month, the chaos had settled into a rhythm. Jonas could answer board questions with conviction because he had finally given himself time to think.

For the first time in years, Jonas ended his week asking not, *Where did my time go?* but *What did my time build?*

From Schedule to State

By now, you have designed the architecture of your ideal week. You have reclaimed your time, rewritten your meetings, and aligned your calendar with your highest leverage. However, there is still a deeper level that we need to address, one that transcends scheduling and enters the realm of experience. You have to consider that it is not about when you work but how you show up when you do.

Flow is not a productivity hack or a zone you stumble into by accident. It is the highest-performing cognitive state available to you as a leader, yet it is often treated as a luxury. It is seen as something for creative professionals, writers, athletes, and others, not executives. That is a costly mistake, since CTOs who scale with grace have one thing in common: They protect their flow state like oxygen.

They build systems that guard their attention, design teams that don't need constant pings, and shape digital and physical environments that reduce noise and increase depth. When you are in flow, you are more creative, more decisive, and more human, and in the complexity of modern leadership, that is not optional but essential. Are you ready to move from managing your week to managing your state?

CHAPTER 13

Flow State Leadership: The Ultimate Productivity

Flow is not a luxury but a leadership state you must protect at all costs.

Studies have shown that people in flow can be more effective in their tasks, but "effective" is an understatement. The data describes a transformation in capability. McKinsey's study documented an extraordinary 500% increase in productivity during flow (Kotler, 2014), while research from the University of Sydney found a 430% boost in creative problem-solving (Searle, 2023). The author further states that even the rate of learning accelerates, with DARPA studies demonstrating a 490% improvement in skill acquisition.

The concept of "flow" was popularized by psychologist Mihaly Csikszentmihalyi, who defined it as a mental state of total immersion in a task: a feeling of energized focus, full involvement, and deep enjoyment in the process (Kotler, 2014). By studying highly skilled individuals ranging from artists to elite athletes, Csikszentmihalyi identified that this state isn't just a pleasant by-product of work but a primary driver of peak performance.

This state is the "holy grail" of technical productivity. Recent research confirms that "both employees and organizations can experience the benefits of flow, leading to improved productivity, creativity, and overall well-being" (*Achieving Peak Performance*, 2024). For a developer, flow is the difference between struggling with a single bug for hours and refactoring an entire subsystem in a single afternoon. When a CTO optimizes for flow, they aren't just improving morale; they are maximizing the cognitive ROI of their most expensive asset: human intelligence.

Here is the nuance for leadership: Your tasks are more than tactical; they are cognitive orchestras that allow you to switch from deep technical problems to financial decisions, from vendor updates to strategic planning. Flow is the space where those transitions become smoother, where insight emerges without effort.

A tool many CTOs use to intentionally access flow is the "flow card," a personal, revisable reference that outlines conditions under which you flow most easily:

- **Time of day:** When are you naturally focused?
- **Location:** What space supports your depth?
- **Inputs:** What preparation do you need?
- **Outputs:** What kind of work best fits your flow mode (writing, reflecting, etc.)?

One CTO's flow card looked like this:

8–11 a.m., closed office door, lo-fi playlist,
handwritten sketchpad, no Slack

He protected this window like a critical system dependency. Within that zone, he worked on architectural road maps, wrote vision docs, and explored unknowns—the kind of work that defined the company's future. Flow is not accidental; it is designed, and it starts with knowing what conditions unlock your best thinking.

REFLECTION PROMPT

Where does your flow state most often get interrupted—people, pings, clutter, or fatigue? Create your first flow card: When are you at your best? What are your triggers for flow? What kind of work belongs in that zone?

Attention Defenders

As mentioned in previous chapters, cognitive science tells us that our brains are not meant to operate in persistent partial attention. Every ping or unscheduled interruption costs more than the moment it takes. It leaves residue: mental fragments that clutter your RAM and throttle clarity. For CTOs, protecting attention requires deliberate systems. These are your *attention defenders*.

Before implementation, many CTOs experience their workday as fragmented, with 10-minute bursts of focus between back-to-back context shifts. They end days mentally exhausted but with a nagging sense that nothing truly moved forward. After designing attention protection, a different pattern emerges: long stretches of focus, fewer escalations, clearer decision-making, and an energized close to the day.

How to start:

- **Layered notification design:** Turn off nonurgent notifications. Use do-not-disturb during deep work blocks.
- **Interrupt protocols:** Create norms with your team about what is urgent. Teach them to self-sort communication: real-time, asynchronous, or batched.
- **Task queue limits:** Work-in-progress limits are not just for Engineering. Apply them to your personal workload. Too many open loops means diluted presence.
- **Cognitive anchoring:** Start each day with a written anchor, which is a sentence about what matters most today. It guides attention through chaos.

In my coaching experience, I have seen CTOs drop context switching by 40% just by having asynchronous update rituals. Instead of real-time check-ins, engineers record two-minute Looms summarizing blockers and progress. The result? Clarity without interruption. Finally, here is a deeper truth: Defending your attention signals what you value, creating personal permission. When you choose focus, your team learns it's okay to choose depth too.

Communication Architecture

The problem I come across regularly is that we don't design communication systems; most of them are inherited, such as a Slack channel, a Notion page, dozens of unstructured updates, and meetings. Before a communication architecture is in place, every piece of information feels urgent. People ping at random, meetings sprawl without purpose, and clarity is rare. But once structure exists, leaders can reclaim time and trust.

Just as you would design a resilient technical architecture, you must design a communication architecture optimized for clarity, scalability, and focus. Start with these layers:

- **Asynchronous first:** Default to written updates. Daily asynchronous check-ins in Slack or Notion. Asynchronous wins back the time you did not know you were losing.
- **Structure synchronous encounters:** Reserve meetings for alignment, tension, and decision-making, and not for information transfer. Every meeting should have a role and a format.
- **Communication cadence:** Weekly team syncs, monthly retros, and quarterly vision alignment. These rhythms reduce the need for ad hoc pings.
- **Routing rules:** Where should what go? Product feedback in Productboard, technical decisions, TDRs, or blockers in Slack #team-name. Remove ambiguity.

One CTO I coached created a simple flowchart for his team: *Where do I put this?* It routed ideas, blockers, and updates to the right place. It eliminated noise and empowered ownership. Before implementing this system, that same team had 25–30 messages per day. Post-implementation, signal-to-noise improved by 60%. Decision latency dropped, and developer satisfaction rose. When you think about communication, build for signal, not volume.

The Digital Environment

Your apps and tools are also not neutral; they can either support or erode your flow. Before revising your tool stacks, it is common to see tabs multiply, platforms distract, and emails drive the day. The average CTO uses 10–15 digital platforms daily. Without discipline, these tools fragment your day into microtasks and your mind into chaos. To redesign your digital environment for flow, use the following techniques:

- **Visual hygiene:** Reduce visual noise by using minimalist themes, collapsing sidebars, and closing unused tabs. Fewer visual stimuli = more focus.
- **Dedicated zones:** Segment tools by the purpose for which they will be used.
- **Tool stack audit:** Review quarterly which tools generate more clarity than clutter, and eliminate redundant or unused platforms.
- **Contextual launchpads:** Create browser bookmarks or desktops configured per mode: coaching mode, writing mode, strategic mode, and so on.

One CTO had a "writing desktop" with only Notion, calendar, and Spotify. When he entered that mode, everything else disappeared. It was a cognitive signal: *You are now doing depth work.* After realigning his digital stack, he cut 2.5 hours of context-shifting per week, which he redirected toward technical vision work. The result was a 12-month architectural plan that had previously eluded him for quarters.

The Physical Space

Imagine sitting down to work in a space that feels calm, clear, and utterly intentional. Not cluttered or chaotic, or a storage closet repurposed into a Zoom dungeon. That is, for most CTOs, the harsh reality. Their space is a catch-all with unread books, stray cables, a half-finished to-do list taped to the wall, and yesterday's coffee still on the desk. The average time-to-flow (the time it takes to move from starting work to entering real focus) stretches longer in a cluttered and noisy environment.

Physical space is programmable, yet most modern offices are programmed for distraction. Harvard Business School research using wearable technology reveals that open offices can reduce productivity by roughly 66% compared to private workspaces, largely by triggering withdrawal rather than collaboration (Camerota, 2018). The toll is also physiological: UC Irvine found that cubicle workers receive 29% more interruptions and suffer 9% higher exhaustion rates than those in private offices (Mark et al., 2008). In fact, the stress is so acute that Scandinavian studies found open-office workers take 62% more sick days (Pejtersen et al., 2011). Architecture is not neutral; it either fuels your team or drains them.

The point is not style; it is signal. Your space should say: *This is where leadership work happens.* After redesigning, CTOs report a reduction in time-to-flow of as much as 70%. They sit down and drop in fast. Your workspace should be treated as infrastructure, and acoustics are a critical component of that stack.

Noise is not just an annoyance; it is also a cognitive leak. Cambridge University research shows that just 8 minutes of open-office noise triggers a 34% increase in stress response and a 25% increase in negative mood (Sander et al., 2021). Crucially for engineers and leaders, this acoustic environment specifically disrupts tasks requiring complex verbal processing and logic. You would not host your database on a server with 25% packet loss; you shouldn't host your brain in an environment with 25% cognitive loss.

The Recovery Protocol

Your mind is not like a central processing unit, but rather like a moving system that thrives with cycles of focus and restoration.

At the same time, CTOs often run themselves like servers with no maintenance windows: always online, always accessible, and always depleted. Burnout rarely announces itself; instead, it arrives subtly with decision fatigue, emotional numbness, loss of vision, and avoidance. By the time it is visible, it's already expensive.

Enter the recovery protocol, which is not a vacation but a repeatable system for cognitive restoration. It might include these elements:

- **Deliberate shutdowns:** Close your laptop. No, really close it. Don't end the day with a half-read tab or a half-drafted Slack message. End clean.
- **Weekend rituals:** One CTO created a rhythm: Friday reflection document, Saturday tech-free walk, and Sunday long-form vision sketching. That cycle fueled clarity.
- **Microrecoveries:** Five-minute breaks every 90 minutes with water, movement, and breath. It is not a hack but a recharge loop.
- **Exercise as fuel:** A daily practice of getting the heart rate up does more than improve health. It enhances executive function, reduces stress hormones, and strengthens resilience. Movement sharpens thought, and the CTOs who treat physical activity as a part of their leadership system and not a hobby consistently sustain higher levels of clarity and endurance.
- **Monthly reviews:** Once a month, zoom out. Ask yourself if you are leading from overflow or from exhaustion.

When you do not take the time to properly recover, the likelihood of burnout and other stress-related syndromes increases. Protect your role and your work-life balance by implementing protocols that will help you rest or become more focused.

Refactor in Action

Aisha adjusted her webcam for the third time that morning. Behind her, the spare room told on her: a pile of laundry, a half-assembled bike, and a whiteboard covered in half-erased equations from three weeks ago. It was the kind of visual clutter that mirrored exactly how she felt inside.

Her days were no better. Meetings bled into Slack pings, which bled into inboxes, which bled into late-night note-taking she barely remembered writing. Focus was something she kept promising herself she would "get back to"—after this release, after this fire drill, after the next quarter.

The breaking point was almost laughable: She spent 45 minutes in a "deep work" block moving files around her desktop, convinced she was "getting organized." By the time she opened the actual document she meant to work on—a critical strategy draft—the time was gone. She didn't feel productive but depleted.

That night, she realized it wasn't just clutter stealing her flow. She never ended a day; she just collapsed into the next one, carrying yesterday's noise into today's work.

So, she rebuilt, not just her space but also her system. The laundry was moved, the bike relocated, and the whiteboard wiped clean. On the desk: only a lamp, a notebook, and one plant. She drew up her flow card—*8:00–11:00 a.m., phone in the drawer, noise-canceling headphones, one task only*—and treated it like a system dependency.

She also created a shutdown ritual. At the end of each day, she wrote three bullet points in her notebook: what she'd finished, what carried forward, and one line of gratitude. Laptop closed. Phone in the drawer. Ten minutes outside, no exceptions.

It wasn't dramatic, but it was enough. Her mornings stopped feeling like a restart of yesterday's chaos. Flow returned—not as an accident, but as something she had designed and defended.

Your Next Frontier: Saying No as Strategy

By now, you have built a week, designed for depth, created conditions for flow, and structured your leadership life around what truly matters. You have optimized your energy, reclaimed your calendar, and reshaped how your organization works around you. At the same time, even the most elegant system can be undone by a single habit: saying yes to the right things.

Every "yes" is a silent "no" to something else: often the work that only you can do. CTOs often underestimate just how unclear their boundaries are. A seemingly small favor for an executive, a quick detour into a project that lacks strategic value, or maybe it is an invitation to a meeting you do not need to attend but feel obligated to. One by one, these "yeses" fill the calendar and drain your focus, until your OS, the one you worked so hard to build, starts running other people's code.

That is the next evolution. In the following chapter, you will explore strategic refusal, which is more than just saying "no"; it is about shaping how others think about your time. You will learn how to use opportunity cost as a compass, develop a decision matrix that filters what truly deserves your attention, and deploy diplomatic "no's" that maintain trust while reinforcing your priorities. Being an effective CTO is about doing the right things, not more things, and protecting your ability to do them. Let's examine that lack of filters and how to build them in the next chapter.

CHAPTER 14

The Strategic No: Protecting What Matters

Great CTOs do not just protect their time; they also shape expectations that seal it.

As a CTO, you likely receive dozens of demands every day. Every request seems small in isolation: a "quick" sync, a "brief" review, or a "simple" favor. But as you have seen by now, each one incurs a hidden cost of time, attention, energy, and strategic displacement. Therefore, it is safe to say that the most dangerous cost for a CTO is not the task itself but what the task displaces.

Every "yes" is a silent "no" to something else, but our brains are wired to ignore this trade-off. Behavioral economics research from Nobel laureate Richard Thaler demonstrates a phenomenon called "opportunity cost neglect": When making decisions, we systematically fail to consider foregone alternatives unless explicitly prompted to do so (*Opportunity Cost Neglect*, 2025). We see the immediate value of the new project, but we are blind to the project we just killed by diluting our resources. This calculator forces that invisible cost into the light, making the "silent no" explicit before you commit.

If you say "yes" to joining a customer call on short notice, what don't you do instead? Maybe it is a draft of the new hiring rubric that could fix your pipeline or an hour of focused writing on next quarter's objectives and key results. Perhaps it is even recovery time that you need to recharge your thinking for a board meeting tomorrow.

It is common to operate in a zero-sum attention economy. That is why I teach CTOs to use an *opportunity cost calculator* as a mindset. Ask these questions every time a request hits your inbox, calendar, or conscience:

- *What will this cost me in focus, energy, and context switching?*
- *What could I do with this time instead?*
- *Am I the only person who can do this, or just the most convenient one?*
- *If I say "yes" now, what will I say "no" to by default?*

One CTO installed a personal rule: If a request costs more than 30 minutes, he would write down what he was giving up. Over time, he saw a pattern that these requests were costing him strategic projects, which were constantly being displaced by reactive favors. That list became a visual cue: a reminder that this best work needed fewer interruptions, not more time.

Opportunity cost awareness is not about being stingy but rather about being sovereign. The default state of a tech leader is *overcommitted*. The strategic CTO is fiercely selective, and when this mind scales across the team, decisions stop piling on your plate. Instead, time becomes intentional, focused, and leveraged.

The Decision Matrix

As you have seen in previous chapters, the CTO's job is not to do the work but to grow the people who do. This requires aggressively protecting the cognitive environment in which those people operate. Just as there is a cost to every decision you make as a CTO, there is a corresponding cost imposed on your development teams that is paid in attention, energy, and focus.

When leaders lack intentional strategies for protecting time and attention, and fail to say no to fragmentation, that cost is transferred directly to the team. Context switching, constant interruptions, and reactive prioritization erode the conditions required for flow. Research shows that without this protection, workers experience a 56% reduction in productivity from app switching alone, with nearly half reporting mental exhaustion from relentless context shifts (Maloney & Puente, 2019).

Once you have embraced the idea that not everything deserves your "yes," the next challenge becomes operational: How do you decide? This is where you would use the well-known and widely adopted decision matrix to work in your favor as a tactical firewall. Its purpose is to quickly sort requests by value and fit. It is your personal algorithm for protecting flow, focus, and impact.

Here is the well-known version of the matrix you can adapt to your needs:

Importance	High urgency	Low urgency
High importance	Do it now	Schedule it
Low importance	Delegate it	Decline it

This 2×2 is just the start, since there is a more nuanced layer that comes from asking: *What category of work is this?* To solve this, use the rule of three filters:

1. **Strategic work:** Moves the company forward in big leaps (such as hiring leaders, defining architecture, shaping culture).
2. **Supportive work:** Keeps others unblocked (including code reviews, vendor decisions, and process design).
3. **Substitutional work:** Things others can or should own (for example, attending every meeting and ad hoc troubleshooting).

Every request falls somewhere, and the dangers of overvaluing support work are that while it feels good and you feel helpful, over time, it will displace your energy.

One CTO decided to organize her agenda to identify the type of work she was performing. After one week, she saw that 65% of her time was substitutional, 25% supportive, and only 10% strategic. The matrix did not just sort decisions; it also sparked a redesign of how she led. Once you systematize your decisions, you depersonalize your "no's" and understand that you are not rejecting people but protecting the work that only you can do. When this happens, you stop being the bottleneck, and your team begins to rise in your absence.

REFLECTION PROMPT

Think of the last five requests you said yes to. Which ones were truly strategic? Which ones displaced something only you could do?

That said, even the most disciplined systems have exceptions. For me, there is one: the CEO. I always take a call from my CEO (regardless of where I am, how I feel, or what's on my calendar). Those conversations are not always urgent in the tactical sense, but they are vital. They are also totally up to you to evaluate and then implement what works best for you.

Sometimes it is about processing a tough decision, sometimes it is venting, and sometimes it is simply needing a thought partner in real time. I index high on being available in those moments because they reinforce trust and cohesion at the very top of the organization.

Of course, this level of openness has trade-offs. It can fracture focus and erode boundaries if left unchecked. But the upside is that when a CEO knows they can count on you, not just for delivery but also for presence, it is worth it. For me, that is not just an exception to the matrix; it is also a principle about how I want to show up in leadership.

Political Capital: Earning the Right to Say No

A strategic "no" is only effective if backed by political capital. Without a track record of delivering tangible value, a refusal risks being interpreted as obstruction or incompetence. Think of political capital as a bank account: You must consistently make deposits by shipping wins, achieving milestones, and earning trust before you can afford to make withdrawals in the form of saying no. The strength of your credibility determines whether your boundaries will be respected or challenged.

When you find yourself low on political capital, or when a request comes from the CEO around a critical business priority, an outright "no" can be dangerous. Instead, consider framing your response in a way that preserves influence while communicating constraints. This is where the "yes, and..." strategy becomes a powerful tool.

For example, you might respond as follows:

> Yes, we can accelerate that feature to secure the Series B milestone. To hit that date, we'll need to defer the mobile refactor. This moves our technical debt "interest payment" to next quarter, but it ensures we hit our immediate growth target.

This approach does not reject the request; it reframes the conversation as a set of trade-offs. You are not blocking the business but highlighting the resource realities that make multiple simultaneous priorities infeasible.

By presenting the trade-offs early, you shift the responsibility from yourself to objective constraints. The CEO or leadership team now has the information they need to make an informed choice. In doing so, you protect your credibility and the organization's outcomes, demonstrating that strategic refusal is not about obstruction but rather about clarity, influence, and enabling better decision-making.

The Diplomatic Arsenal

Saying "no" is hard. It is even more difficult to say it while maintaining trust, respect, and alignment. Doing so is more practice than art. CTOs who cannot diplomatically say no end up either resentful or burned-out. The trick is to refuse without rejection and preserve relationships while reinforcing your priorities.

The ability to say no is more than a tactical maneuver: It is a psychological lever that shapes your authority, clarity, and resilience as a CTO. Research in organizational psychology demonstrates that leaders who clearly exercise their power without intimidation can prevent cognitive overload in their teams. When CTOs fail to say no, they unintentionally encourage a culture of dependency and overreach, where team members hesitate to take initiative, defer decisions, or withhold critical input. Conversely, a respectful refusal can empower teams to solve problems independently, strengthening both competence and confidence.

The act of saying "no" is intrinsically tied to boundary-setting and self-care. Wei (2024) states that asserting limits improves mental health, reduces stress, and increases self-esteem. Leaders who habitually overcommit experience a constant drain on cognitive and emotional resources, which over time leads to burnout. By contrast, strategic refusals create psychological breathing room, allowing the CTO to focus on high-leverage decisions and long-term strategy rather than being consumed by reactive demands.

I usually coach leaders to develop a diplomatic arsenal, which is essentially a repertoire of language tools that soften a boundary without softening the principle. A few go-to patterns include the following:

- **Deferral:** "This is important, and I want to give it real attention. Can we revisit next week when I am out of delivery mode?"
- **Reframe:** "This looks like something better owned by [person]. Let's empower them to take it and circle back if they need support."
- **Trade:** "I can take this on, but I'd need to delay [what would be delayed]. Which one is more urgent?"

- **Policy:** "As a rule, I do not join first-round vendor calls unless we have already scoped a technical requirement."
- **Appreciation and decline:** "Thanks for thinking about me. I am at capacity right now and don't want to do this halfway. Let's find another path."

These practiced responses are clear and objective, allowing you to practice and apply whenever necessary. Before developing this muscle, it is common to default to one of two extremes: avoidant "no's" that feel cold, or overextended "yeses" that backfire. After refining their diplomatic arsenal, leaders report higher trust, better boundaries, and more respect for their time from executives and their own teams.

The real shift, looking closely, is not how you will decline but how you use the model decision hygiene to protect more than your calendar. Using this model, you will train others how to engage with your time and make the best use of it. You will notice that the power of no is a cornerstone of sustainable leadership. When used thoughtfully, it protects your mental energy, empowers your team, reinforces trust, and magnifies your strategic impact. Saying no is not denying responsibility; it is creating an affirmation of purpose and a tool for creating focus, alignment, and long-term resilience.

The Expectations Manager

Expectations are built from patterns: If you always say yes, people will expect yes. If you always deliver instantly, people will expect speed over quality. Finally, if you do not set boundaries, others will set them for you.

The expectations manager is the inner operating system of strategic leadership. It is more than avoiding responsibility; it is about shaping what your team and your stakeholders expect from you and each other. A CTO once told me that his calendar was full of emergencies that he trained people to bring to him. That admission changed everything.

Start by auditing your "yes" patterns: Look at the last month and identify what you have said yes to that someone else could have done, decided, or deferred. Then, map the expectations your behavior created. Ask yourself if others expect you to make decisions, every sync, and every bug fix, or if you are letting them decide on their own.

Lastly, reset the defaults by announcing a new policy. Something along the lines of, "My calendar is now optimized for strategic work. That means less availability for ad hoc syncs. Let's work through async or go through leads." Remember that expectations are habits, not mandates, which can be reshaped through consistent modeling.

Once you do this, you create something remarkable: space.

Space for others to lead, for issues to self-resolve, and for strategic drift to be noticed before it is irreversible. The strategic no is more than a boundary; it is a leadership signal that tells the team, "I trust you to run without me, and I will invest my time where it unlocks the most."

The Counteroffer

Sometimes, a flat "no" is not the answer, but neither is a passive "yes." The counteroffer is a strategic move to help you move the conversation from obligation to collaboration. Instead of absorbing the request, you redesign it. Here are some examples you can use:

- **Request:** "Can you review this document today?"
  - **Counteroffer:** "I can review it by Thursday. If it is urgent, can [person] take a first pass?"

- **Request:** "We need you in this decision."
    - **Counteroffer:** "I can clarify the decision criteria and trust you to make the call. Loop me in only if it hits red flags."
- **Request:** "Can you present this to the board?"
    - **Counteroffer:** "How about I coach you to present it? That way, the board starts to see your leadership first-hand."

The counteroffer moves you from executor to enabler and allows you the chance to delegate without disengaging, build leadership in others, teach decision frameworks, and preserve your bandwidth for higher-leverage work. This is how the CTO role scales with redesign and not through accumulation. Once you start being the center of gravity, you become a capacity architect, deciding what you can and cannot take on.

Refactor in Action

Arjun had a reputation for being the "yes" CTO. Need someone to sit in on a vendor call? Yes. A last-minute customer demo? Yes. Debug a production issue at 11 p.m.? Always yes.

At first, it felt good—like leadership. He was visible, responsive, and indispensable. But over time, his calendar filled with other people's priorities, and his own work slipped to the margins.

The wake-up call didn't come from burnout or a missed board deadline. It came while at home. One Friday night, his eight-year-old daughter proudly handed him a crayon drawing from school. It showed a stick figure hunched over a glowing rectangle. Above it, in crooked letters: *My dad builds emails.*

Arjun laughed at first—until he saw her face. She wasn't joking. To her, that was what he did. Not building products. Not leading teams. Emails.

The following Monday, he started small. Instead of reflexively saying yes, he paused. When asked to join yet another Product sync, he replied, "Loop me in if it crosses X threshold." It felt awkward, almost rude. But then he tried again. A colleague asked him to review a 20-page deck that day. He countered, "I can review by Thursday. If it's urgent, can someone else take the first pass?" The world didn't fall apart.

It wasn't smooth. A few people bristled. Some work slipped. But he began to notice something different in himself: For the first time in months, his desk wasn't buried in other people's emergencies.

The next morning, his daughter's drawing was still pinned to the fridge.

Arjun stood staring at it longer than he meant to, laptop buzzing in his hand. He was late for yet another "quick sync." This time, he didn't rush.

Your Next Frontier

Mastering the strategic no is more than not saying yes; it is about making space for what truly matters. You have trimmed the noise, set new boundaries, and modeled better decision-making. *What is still missing?* Consider clarity and selectivity only half of the game. Now comes in the deeper work: designing your leadership for the long run.

While protecting your calendar is one thing, protecting your *capacity* is another. If you don't establish rhythms that replenish your energy, connect you to your purpose, and prevent silent drift, no amount of strategic clarity will save you from burnout.

To settle this, in the next and final chapter, you will explore how great CTOs don't just optimize for performance; they engineer for sustainability by working in patterns that fuel longevity, presence, and impact. Chapter 15 is where we stop sprinting and start building the systems that let you play and win the infinite game.

CHAPTER 15

Sustainable Rhythms: Playing the Infinite Game

Stability is not slowness;
it is the speed you can maintain forever.

The path from developer to CTO is paved with the currency of output: shipping features, killing bugs, and scaling infrastructure. In the early stages, your value is a direct reflection of your personal throughput. But as the organization grows, the linear math of "grind" begins to fail. You eventually reach a threshold where your busyness is indistinguishable from stagnation, and your adrenaline is merely a mask for structural fragility. At this level, a leadership style that runs on heroic effort is a leadership that is running out of time.

To cross this chasm, you must stop measuring success by what crosses your desk and start measuring it by the capacity you build. True leadership leverage is an intentional architecture, built layer by layer: first through systems that scale judgment, then through people who multiply impact, and finally through decisions that define the future. This book is about that transition—moving from the "central processor" of the company to its most powerful catalyst.

This final chapter is about what happens next: building sustainable rhythms that preserve your energy, sharpen your presence, and make leadership feel like a practice and not a race. Leadership is not a sprint, nor even a marathon—it is an infinite game where the goal is to keep going.

Before CTOs learn this, their calendars might look impressive, but their recovery time is nonexistent. They crash after board meetings, miss their loved ones' special events, and avoid vacations because it takes too long to reenter their mental stack. Even when attending meetings in person, they are in "present in body, absent in mind" mode. The most strategic CTOs think differently: They treat their schedule like a heartbeat, with expansion, contraction, tension, and release. They systemize those cycles so their leadership is not just effective but also renewable.

You need patterns that hold you, not more hustle. It is time to start designing your rhythm to make it work for, and not against, you.

The Review Rituals

Rhythm is how you reflect your productivity. Review rituals are not indulgent; they are infrastructure. Without them, you operate from momentum, not intention, you automatically become reactive, not rhythmic, and your leadership becomes brittle. The CTOs who sustain long arcs of performance consistently do one thing: They review. Not just tasks, but also themselves, their impact, their alignment, and their drift. Below is how they do it.

Daily Wrap-Ups

Use five minutes at the end of the day to ask yourself three questions: *What did I move forward? Where did I drift?* and *What deserves tomorrow's focus?* More than journaling, this is mental closure, and you are writing the commit message for your day.

Weekly Resets

On Friday afternoons or Monday mornings, use a one-hour recurring block of time to reflect on what the ratio of strategic to reactive work was, where you said yes instead of no, and the one decision you need to make before it makes itself. This is your *leadership stand-up*, meaning not just what you are doing but also how you are doing it.

Monthly Reflections

Every month, take 90 minutes to combine metrics, mood, and narrative. Ask, *What changed in my leadership this month? What surprised me? What needs to be redesigned?* Look at your calendar and identify more than what is there; *what is missing?*

Quarterly Retreats

Take a half-day, every three months, to leave your usual work zone and reflect on energy, focus, culture, and relationships. This time is not about goals but about state and system health. Great CTOs don't just review their organizations; they also review themselves as systems.

At the same time, rituals only work if they are honored, so make them visible, block the time, and build the templates. Do not wait until you are burned-out to start reflecting, as that is similar to patching a server during an outage.

One CTO I coached began with a 10-minute journaling habit each evening. Six weeks later, he reported fewer leadership regrets, faster prioritization, and a sharper ability to name hard trade-offs. "It is like I recompiled my day before shutting down," he told me. He understood that reflection is how to learn from experience before it becomes pain.

Energy Management

Energy is the currency *beneath* the currency, and protecting it requires understanding exactly what drains it. As you saw in Chapter 2, Christina Maslach's renowned research at UC Berkeley identifies six distinct drivers of burnout: workload mismatch, lack of control, insufficient rewards, breakdown of community, absence of fairness, and value conflicts (Maslach & Leiter, 2022).

A sustainable rhythm is not just about "rest"; it is also an architectural fix for these six vulnerabilities. It addresses workload through boundaries, control through autonomy, and values through life integration, ensuring your energy expenditure aligns with your human needs rather than fighting against them. You can have a full day of open hours and still be useless if your cognitive battery is drained. So, the question becomes, *How do I engineer for energy?*

Start with your energy map—a simple but brutally honest tool:

- **Step 1—Track for a week:** Every 2–3 hours, log your energy from 1 to 5. Note what activity you were doing and highlight what drained vs. what restored you.
- **Step 2—Spot patterns:** Ask yourself what drains you and what sparks you. Are you always depleted after 1:1s? Do you get a spike from systems thinking? Is your morning your prime time or your fog zone?
- **Step 3—Redesign your week:** Move strategic work to high-energy windows and group energy-draining tasks together, with recovery buffers. Block your peak times for high-impact, high-focus work only.

CTOs who deliberately manage energy see surprising gains. One leader I worked with swapped all heavy decision meetings from late afternoon to late morning. The result? Fewer escalations, better notes, and fewer reworks. Another CTO changed his exercise routine from evening to early morning. "It flipped my whole output curve," he said. "Suddenly, I was catching up, setting the pace."

Remember that energy management is also about inputs, not just protection. Ask, *What stabilizes my clarity? What social interactions boost my morale?* and *What rest practices actually restore me, not just distract me?*

Finally, protect your recharge time like a production system. A CTO once said, "I used to feel guilty for taking Fridays off. Now, I realize my Saturday self is what my team works with on Mondays." Your output is only as good as your energy input.

The Connection Balance

Leadership is more than a solo act, but that does not mean it should be a constant chorus. Connection is one of the most mismanaged resources in a CTO's tool kit. Not because we do not value it, but because we often confuse presence with proximity. In other words, being constantly available, responding instantly, and jumping into every Slack thread as a performance of leadership.

The paradox is that the moment you insert yourself into the stream of interactions, the more you erode your ability to think, discern, and lead from altitude. The solution, however, is not withdrawal. You must use intentional connection to interact with your team effectively.

When CTOs learn to calibrate their connection bandwidth, they stop confusing busyness with influence. They begin to treat relationship-building as a lever, not a liability. Their time becomes more about trust than traffic. However, for this to happen, you must start rethinking the default. Before reengineering the connection, most CTOs fall into one of two traps, neither of which is sustainable:

- **Always-on operator:** They reply to everything, they are in every meeting, and their badge of honor is being "the most responsive person in the organization."
- **Ghost leader:** After overcommitting and burning out, they disappear behind a wall of Slack silence and unapproachable calendars.

What sustainable CTOs design instead is asymmetric visibility. Their presence is both deeply felt and not constantly performed. Think of your relationships across three concentric circles: *core* (CEO, cofounders, and functional C-level peers), *strategic layer* (tech leads, senior individual contributors, and product managers), and *organizational layer* (broader engineering and cross-functional teams). Each level requires different rhythms, expectations, and touchpoints.

For your core, weekly syncs and deep dives are nonnegotiable. These are your sources of alignment where you can navigate ambiguity. For the strategic layer, your goal is empowerment. Biweekly 1:1s, asynchronous updates, and periodic check-ins matter more than real-time guidance. You let them lead as you coach the system. Finally, for the organization layer, you have to symbolically show up. This can be done with company-wide updates, town halls, shared documents, and vision memos. Here, visibility will create clarity, not intimacy.

REFLECTION PROMPT

What's one connection you've been maintaining out of habit, not value? What would full integration look like—not in theory, but in one calendar block this week?

One CTO we coached redefined his entire connection strategy around three types of signals: relational, with deep, 1:1 time that builds trust; symbolic, with company-wide presence that reinforces leadership vision; and informational, where updates and insights are clearly shared, without the noise of meetings. He shifted from 30+ hours of weekly meetings to 4 hours of high-leverage 1:1s, a Friday vision memo, biweekly ask-me-anything sessions, and asynchronous strategy updates via Loom.

The result was that he was in fewer conversations, but more conversations started with his thinking. This is the crux of sustainable connection: Stop aiming to be involved in every decision. Instead, build a communication architecture that makes your thinking accessible, your leadership visible, and your availability predictable.

The Warning System

Burnout starts with whispers, subtle signs that go unnoticed until they calcify into crisis, not with a great breakdown. Most CTOs are trained to tolerate discomfort, including normalizing long hours, endless decisions, and emotional labor. However, when small misalignments accumulate unchecked, you are not just tired; you are also drifting.

Burnout rarely announces itself; it creeps in through the cracks of a busy schedule. The McKinsey Health Institute's research reveals that middle managers experience the highest burnout rates of all (Brassey et al., 2023). The warning signs manifest as what researchers describe as severe burnout due to overwhelming demands, limited resources, and a lack of strategic support. These are the exact conditions a CTO faces without intentional rhythm design. If you do not build a warning system, the silence you hear is not stability; it is the prelude to a crash.

A personal warning system is not a weakness but observability. It is the logging and the alerting that allow you to reboot before a full system crash. Here is a model suggestion you can use to start building your own warning system:

- **Layer 1—The feel layer:** Start with emotional observability. This is more than just reflecting on *Am I happy?* It is about feeling proud of how you led that week, if you felt defensive or avoidant in any moment, and identifying the topics or people that drained you.
- **Layer 2—The behavior layer:** Notice recurring patterns such as being late to 1:1s you used to enjoy, avoiding feedback conversations, and rereading the same document three times and still missing key details. These are not just productivity dips; they are also warning flares. One CTO noticed he was increasingly using next week as a placeholder for decisions. It became a catch-all for avoidance. That signal triggered a deeper retrospective: He was overstretched and underrecovered. He needed better leadership boundaries, not a vacation.
- **Layer 3—The system layer:** Now, zoom out and review your system: Does your calendar reflect your priorities? Are your meetings mostly outputs or outcomes? Are you

protecting time for recovery? Are you designing your days or defending them? Try implementing a simple Friday check-in, where you ask: *What drained me this week? What energized me?* and *What will I do differently next week?*

Build these layers into your rituals, make them visible, and share them with a trusted peer or coach. The best CTOs don't just monitor their teams; they also monitor themselves with rigor, grace, and precision. Early detection is the difference between course correction and collapse.

The Life Integration

If there is one thing I have seen while mentoring professionals, it is that *you cannot lead sustainably if your life is disintegrating.* Too many CTOs have a public system and a private crisis. Their calendars are pristine and their dashboards green. Hidden, outside the office, their relationships are strained, their bodies depleted, and their inner life increasingly silent. Life integration is not about the perfect balance. It is about alignment, about making sure that your leadership is an extension of your values, not an escape from them.

One CTO blocked school pickups on Tuesdays and Thursdays. Another scheduled *lunch walks* with their partner. A third turned their weekly team update into a practice of gratitude, celebrating effort, not just output. This is all about coherence, not productivity. Here is how a life integration exercise works:

1. List your top five values, and write them down.
2. Map your current week, and identify the percentage of your time that reinforces those values.
3. Spot the dissonance, and mark where you are out of sync.

4. Make one tiny, tangible change. Book the therapy session, reclaim dinnertime, block a solo walk, or do whatever you feel is necessary.

This is called rhythm reinforcement, where you practice integration not as a one-time fix but as a design principle. When your life and leadership support each other, you stop leaking energy and start living wholeness, not just performing it.

Refactor in Action

Samir asked his daughter how her season was going. She shrugged. "You'd know if you came."

Samir blinked. He had been at every investor call, every sprint review, every late-night incident—but not one soccer match.

The next morning, he opened his calendar. It was time-blocked, color-coded, and impressive. But nowhere in it was his family. Nowhere was recovery. Nowhere was he.

That weekend, his wife sat across from him at breakfast, hands wrapped around her coffee mug. "You're here," she said softly, "but you're not *with* us."

It landed harder than any board critique. She wasn't angry. She was tired. Tired of the kids asking if Dad would be working again tonight. Tired of waiting for the man who always promised he'd slow down after this quarter.

On Monday, Samir made a different kind of decision. He blocked Tuesday and Thursday evenings as nonnegotiable family time. He started walking his daughter to practice once a week. He and his wife scheduled a standing Friday lunch, even if it meant moving a leadership sync.

For once, the space in his calendar wasn't meant to be filled. It was a choice he was proud of.

From Rhythms to Road Maps

By now, you have built more than just habits; you have begun to architect a new operating system. One grounded in rhythm, not reaction; clarity, not chaos. You have seen how leverage, time design, and flow are not soft perks but leadership infrastructure. You have also seen that sustainable leadership is not about slowing down but about lasting long enough to matter.

However, this book was not meant to be a collection of tactics. It was always supposed to be a transformation blueprint, which only becomes real when it is lived. So, as we close, we will bring it all together—your roles, your rhythm, and your rules—into one integrated system. The architecture is now yours, the leverage is built, and the time is reclaimed. The only question left is: What will you do with your freedom?

Your Refactored Life

If you have made it this far, you already know this book was not about tips and tricks; it was about systems. You have explored how to design your role so that you do not drown in it. You have learned that leverage is not about doing more yourself but about creating multipliers that extend beyond you. You have seen that time is not merely about hours in the day but about protecting the attention, energy, and focus that define whether those hours matter.

These three pillars—role design, leverage, and time mastery—are not separate skills. They are a single architecture. Together, they form the operating system of your leadership. Without role design, you remain stuck in hero mode, carrying everything yourself. Without leverage, your team waits on you, and your organization stalls. Without time mastery, even the best-designed systems collapse under the weight of distraction. However, when all three integrate, you unlock something rare: leadership that is scalable, sustainable, and deeply human.

The lesson is simple but not easy: You do not need to do more; you need to lead differently. Knowledge is inert until you put it to work. This book is not meant to be admired on a shelf; it is meant to be used. Your implementation path doesn't need to start with a sweeping change. It begins with a single commitment: to treat your role as something that can be designed, not endured.

Start small. Choose one principle from role design and put it into practice. Maybe that's clarifying a decision with your team. Maybe it's writing your first entry in the CTO Notebook. Maybe it's blocking one sacred hour on your calendar this week. Next, add leverage. Ask yourself: *What am I doing today that someone else could own tomorrow?* Don't just delegate tasks; delegate thinking. Invest in your people until they no longer need you for the work you once tightly clutched.

Then, reclaim time. Protect focus as though it were capital, because it is. Audit your week, name your distractions, and remove at least one. The implementation path is a series of deliberate steps, not a sprint, and each one compounds into the next.

Embracing the CTO Manifesto

Before you begin measuring progress, it helps to anchor your actions in a broader philosophy: the CTO Manifesto, available at ctomanifesto.org (de Bruin & Cotton, 2025). Developed by 7CTOs, a community of seasoned CTOs, the manifesto distills the mindset, principles, and behaviors that define high-impact technology leadership. It reminds us that being a CTO is not about doing everything yourself but about shaping systems, enabling teams, and sustaining long-term organizational impact. The manifesto emphasizes seven guiding principles that serve as touchstones for every decision you make:

1. **Business alignment:** Prioritize initiatives that directly advance company goals over personal technical preferences or transient trends.
2. **Financial discipline:** Ensure that technology decisions maximize return on investment, avoid waste, and support sustainable growth.

3. **Team execution:** Focus on empowering your team to make decisions and deliver outcomes independently.
4. **Adaptability:** Be willing to adjust plans and approaches as new information emerges, resisting rigidity in the face of complexity.
5. **Integrity and accountability:** Build trust through transparency, responsible decision-making, and consistent follow-through.
6. **Sustainable innovation:** Foster long-term, repeatable processes for technical and organizational innovation without burning out yourself or your team.
7. **Curiosity and joy:** Prioritize learning, exploration, and psychological safety to sustain engagement, creativity, and motivation.

Instead of delegating tasks, delegate thinking so that you can reclaim your time. Protect focus as though it were capital, because it is. Embracing the CTO Manifesto provides a framework for making intentional decisions, cultivating autonomy in your teams, and structuring your attention so that your influence scales across the organization. When you act on its guidance, each choice becomes a lever that multiplies your impact, setting the stage for the next chapter.

Progress Markers and Obstacles

If you have ever played an arcade game, you know the first screen well: "Insert Coin." Nothing happens until you invest. That is where you are now, and reading this book is not the game but the coin. The game begins when you act.

Your progress will not be measured in how many chapters you marked; it will be measured in what shifts inside your week, your team, and your company. Early markers might feel small:

- Nothing breaks when you stop attending a meeting you used to control.
- A team lead surprises you with a decision you did not have to make.
- You notice an afternoon of uninterrupted focus, and it feels like oxygen.

Later, the marks get bigger, and you will start to see strategic sessions that once required you now running without you. Investors remark not on the technical detail but on your clarity of vision, your organization moves faster, and, paradoxically, you feel calmer. Eventually, you will reach the "Press Any Key" stage. The system works, the team hums, and you have finally understood that it was up to you to build the architecture, apply the leverage, and master your time. Now, you have to decide what comes next.

At the same time, it would be dishonest to pretend the path ahead is smooth. You will hit resistance, and some of the most common obstacles include

- old habits, such as saying yes too quickly, jumping into the weeds, and responding before reflecting.
- organizational expectations, including teams accustomed to instant access that will struggle when you create boundaries, and peers, who may mistake your focus for withdrawal.

- the myth of indispensability, when a quiet voice in your head will whisper that if you do not do something, everything will fall apart (that voice is lying).
- short-term discomfort where delegation feels slow at first, focus feels selfish, and redesign feels disruptive. The temptation to revert will be strong.

But you already know the alternative: Staying on the old path means that you will stay stuck in the overcommitment loop, underleveraged, and forever reactive. When obstacles appear, remember that friction is proof that you are changing the system and that you should not retreat. Adjust, communicate, and continue.

Small Shifts, Big Results

The most underrated truth of leadership is that transformation is rarely dramatic; rather, it is incremental. One sacred block of time, consistently defended, can lead to breakthroughs in strategy. One clarified decision can unlock weeks of autonomy. One documented principle can prevent years of cultural drift. These changes compound like interest on capital, growing invisibly until they are suddenly undeniable.

Look back six months from now, and you won't remember the moment you shifted. You will just notice that you breathe easier, your team decides faster, and your company moves with more coherence. These are the steady results of the compound effect, and steady is what wins the infinite game.

There is one theme that threads through every chapter of this book: Leadership is never done. Leadership might function well at a certain level of complexity until the next phase shift hits you, and the system you built needs to be refactored again. The systems you design today will not serve you indefinitely, as teams will evolve, markets shift, and tools change. The demands on you will look different a year from now.

That is not failure; that is reality. The point of building architecture, leverage, and time mastery is not to freeze the system but to make it adaptable. Think of yourself less as a leader who has arrived and more as a leader who is continuously iterating. Each refactor makes you more resilient, not more perfect. This is the infinite game where you do not win: You play it better and better every time.

Here is where I end and you begin. You now hold the architecture, leverage, and time to lead on your terms. You are no longer trapped by the myth of being indispensable, the chaos of distraction, or the pull of shallow work. You have the tools to reclaim your role, your focus, and your impact.

What will you do with your newfound freedom? Will you use it to push your company toward a greater scale? To invest in your people until they surpass you? To explore ideas you have never had time for? To finally find balance with the life you have built outside of work?

Whatever the answer, remember that freedom is not the absence of responsibility. It is the ability to choose where to place it. The rest is execution, and the time to start is now.

If this book has helped you, share it with another leader who needs it. The challenges you face are not yours alone; they are universal in our profession. Helping another CTO reclaim their architecture, leverage, and time is itself an act of leadership. And if you want to go deeper, revisit your notes. Highlight one principle, one exercise, one reflection prompt. Act on it this week. Then, act on another next week. Change happens not in intention but in iteration.

The coin is already inserted. The screen is waiting.

Press any key to continue.

Your Next Step

Find the Right Room

The CTO role is lonely by design.

You carry strategic responsibility, technical depth, and organizational tension — often without a true peer group.

7CTOs is where experienced technology leaders pressure-test decisions, compare patterns, and refine how they operate.

Inside, you'll:

- Work through real leadership challenges
- Learn from operators who've scaled before
- Access frameworks and vetted advisors that reduce friction

If you're serious about operating at a higher level, this is your room.

Scan the QR code or visit:
https://7ctos.com/community

Acknowledgments

No book is built alone, and this one is no exception. The ideas in these pages were shaped by people who challenged me, believed in me, and walked alongside me through some of the most defining seasons of my life.

Dan Stoneman — Dan, you have modeled for me what it looks like to be a relentless student of human behavior while serving as CIO of one of the largest school districts in the country. You showed me that leadership at scale and genuine curiosity about people are not opposing forces—they are the same force. And you helped me unearth my favorite word in the English language: explore. That word lives in everything I do now, and it started with you.

Dan Martell — Dan, you wear your heart on your sleeve in a world that often tells leaders to hide it. You showed me what it looks like to live your values without compromise—not as a slogan, but as a daily practice. Your book *Buy Back Your Time* changed how I think about the most precious resource any leader has, and your example changed how I live it.

Aaron Contorer — Aaron, you have this rare gift of always meeting me exactly where I am and then, in the most gentle and loving way possible, raising the bar. You challenge my thinking while affirming my being. That combination is rarer than most people realize, and I am a better leader and a better person because of it.

Casey Kleindienst — Casey, you showed me what the role of a boss really is: to bring out the best in people. When I arrived in the United States in 2000, I could never have dreamed what a profound impact you were going to have on the way I see the world. Everything I know about leading people started with what you modeled for me.

Beth Carr — Beth, you supported me through one of the hardest phases of my life. And you were right there with me—in full belief—when the Press Any Key to Continue concept was birthed. You saw it before it had shape, and you believed in it anyway. That kind of faith in another person's vision is a gift I will never take for granted.

Calvin Hendryx-Parker — Calvin, you are a force of joy and optimism in my life. You are the prototype for what a member of 7CTOs looks like—generous, curious, and fully alive in your craft. I will never forget the moment you teared up when you were awarded the Press Any Key to Continue Award at our 0111 CTO Conference. That moment told me everything I needed to know about what we are building.

Brittany Cotton — Brittany, my partner in all things 7CTOs and Coach EXE. God smiled on me the day He put you in my life. I know that when He called Michael home to be with Him, He sent me you. I used to think 7CTOs was mine. But it's yours.

Marissa His — Marissa, you are my partner in organizing my life, and without you, I would be a mess—probably never getting to the important stuff. Thank you for the finesse and elegance with which you manage my time, my feedback, and my dreams. Thank you for guarding my time with your life.

Darnah Thompson — Darnah, you helped me get this book out of my brain and into the hands of my readers. Book 3 is in the bag, Darnah. Let's do two more.

To the family system that keeps me going—because no mission survives without a foundation of love.

Kath — My wife, my anchor, the person who makes everything else possible.

Jacques, Genevieve, and Colette — You are my reason for building something that lasts.

Leon and Esme — Mom and Dad, having you finally under my roof is one of the great joys of my life. Thank you for everything you built so I could build.

To my two lovely sisters, **Ilse** and **Diedre de Bruin**, who have never stopped supporting me—not once, not ever.

And to the **Oelrich clan** — For 30 years of warmth and acceptance that any son-in-law craves. Yes, even you, Manni. I couldn't do life without the support you have given me.

Thank you all. This book exists because each of you, in your own way, pressed a key that helped me continue.

References

Achieving peak performance through flow in the workplace. (2024, August 18). Psychology Town. https://psychology.town/applied-positive-psychology/achieving-peak-performance-workplace-flow/

Age and tenure in the C-Suite: Korn Ferry study reveals trends by title and Industry. (2020). Korn Ferry. https://d1io3yog0oux5.cloudfront.net/_503a0409c6a1028a0ea93cad6bda6e49/kornferry/news/2020-01-21_Age_and_Tenure_in_the_C_Suite_Korn_Ferry_Study_34.pdf

BBC News. (2018, October 10). *Amazon scrapped "sexist AI" tool.* https://www.bbc.com/news/technology-45809919

Brassey, J., Herbig, B., Jeffery, B., & Ungerman, D. (2023, November 2). *Reframing employee health: Moving beyond burnout to achieving physical, mental, social, and spiritual health.* McKinsey. https://www.mckinsey.com/mhi/our-insights/reframing-employee-health-moving-beyond-burnout-to-holistic-health

Camerota, C. (2018, July 9). *The unintended effects of open office space.* Harvard Business School. https://www.hbs.edu/news/articles/Pages/bernstein-open-offices.aspx

Cecchi-Dimeglio, P. (2023, August 29). *How leaders expertly harness the power of "no" to shape success*. Forbes. https://www.forbes.com/sites/paolacecchi-dimeglio/2023/08/29/how-leaders-expertly-harness-the-power-of-no-to-shape-success/

de Bruin, E., & Cotton, B. (2025). *CTO Manifesto*. CTO Manifesto. https://ctomanifesto.org/

Desai, M. S., & Chulkov, D. V. (2011). Escalation of commitment in MIS projects: A meta-analysis. *International Journal of Management & Information Systems (IJMIS), 13*(2), 29. https://doi.org/10.19030/ijmis.v13i2.4944

Developers want more, more, more: The 2024 results from Stack Overflow's Annual Developer Survey. (2025, January 1). *Stack Overflow*. https://stackoverflow.blog/2025/01/01/developers-want-more-more-more-the-2024-results-from-stack-overflow-s-annual-developer-survey/

The Eisenhower Matrix. (2024). Columbia University. https://sps.columbia.edu/sites/default/files/2023-08/Eisenhower%20Matrix.pdf

Enterprise generative AI 2026: The adoption crisis, ROI reality, and strategic imperative. (2025, November 27). Axis Intelligence. https://axis-intelligence.com/enterprise-generative-ai-2026-guide/

Falzone, T., & Treynor Sloss, B. (2024, December 16). *The Evolution of SRE at Google*. USENIX. https://www.usenix.org/publications/loginonline/evolution-sre-google

Gartner predicts 30% of generative AI projects will be abandoned after proof of concept by end of 2025. (2024, July 29).

Scribd. https://es.scribd.com/document/889765289/Gartner-article

Glaveski, S. (2021, May 3). *How to increase productivity by 500% and boost innovation*. Steve Glaveski. https://medium.com/steveglaveski/how-to-increase-productivity-by-500-and-boost-innovation-245e0df951ae

Goltz, S. M. (2000). Escalation research: Providing new frontiers for applying behavior analysis to organizational behavior. *The Behavior Analyst, 23*(2), 203–218. https://doi.org/10.1007/bf03392011

Grove, A. S. (2015). *High output management*. Vintage.

Harvey, N. (2026, January 5). *A history of DORA's software delivery metrics*. DORA. https://dora.dev/guides/dora-metrics/history/

Holland, J. H. (2006). Studying complex adaptive systems. *Journal of Systems Science and Complexity, 19*(1), 1–8. https://doi.org/10.1007/s11424-006-0001-z

Jain, S. (2025, June 20). *How SRE revolution is changing software engineering*. Medium. https://medium.com/@sajal.devops/how-sre-revolution-is-changing-software-engineering-487ccb3a8f1d

Kalliamvakou, E. (2021, May 25). Octoverse spotlight 2021: The good day project—personal analytics to make your work days better. *The GitHub Blog*. https://github.blog/news-insights/research/octoverse-spotlight-good-day-project/

Kalliamvakou, E. (2022, September 7). Research: quantifying GitHub Copilot's impact on developer productivity and happiness. *The GitHub Blog*. https://github.blog/

news-insights/research/research-quantifying-github-copilots-impact-on-developer-productivity-and-happiness/

Keil, M., Depledge, G., & Rai, A. (2007). Escalation: The role of problem recognition and cognitive bias. *Decision Sciences, 38*(3), 391–421. https://doi.org/10.1111/j.1540-5915.2007.00164.x

Keil, M., Mann, J., & Rai, A. (2000). Why software projects escalate: An empirical analysis and test of four theoretical models. *MIS Quarterly, 24*(4), 631. https://doi.org/10.2307/3250950

Kotler, S. (2014, January 8). Is the secret to ultimate human performance the F-word? *Forbes*. https://www.forbes.com/sites/stevenkotler/2014/01/08/the-research-is-in-a-four-letter-word-that-starts-with-f-is-the-real-secret-to-ultimate-human-performance/

Maloney, D., & Puente, L. (2019, October 31). *The app era is here: Now to optimize for it*. Slack. https://slack.com/intl/es-es/blog/transformation/the-app-era-is-here

Mark, G. (2023). *Attention Span: A Groundbreaking Way to Restore Balance, Happiness and Productivity*. Hanover Square Press.

Mark, G., Gudith, D., & Klocke, U. (2008). *The cost of interrupted work: More speed and stress*. https://ics.uci.edu/~gmark/chi08-mark.pdf

Martell, D. (2023). *Buy Back Your Time*. Penguin.

Maslach, C., & Leiter, M. P. (2022). *The Burnout Challenge*. Harvard University Press.

Mathebula, B., & Barnard, B. (2020). *The factors of delegation success: accountability, compliance and work quality.* ResearchGate. https://www.researchgate.net/publication/339747499_The_factors_of_delegation_success_accountability_compliance_and_work_quality

Moore, K. (2023, June 19). The power of saying no. *Psychology Today.* https://www.psychologytoday.com/us/blog/mind-matters-from-menninger/202111/the-power-of-saying-no

Murty, R. N., Dadlani, S., & Das, R. B. (2022, August 29). *How much time and energy do we waste toggling between applications?* Harvard Business Review. https://hbr.org/2022/08/how-much-time-and-energy-do-we-waste-toggling-between-applications

Nawrat, A. (2021, June 9). *Study: Too many tech tools are killing remote workers' productivity.* Unleash. https://www.unleash.ai/hr-technology/qatalog-tech-killing-productivity-remote-workers/

Nikos-Rose, K. M. (2023, April 13). *UC Davis study finds tweets can amplify, disrupt, unite and divide.* UC Davis. https://www.ucdavis.edu/news/tweets-can-amplify-disrupt-unite-divide

Opportunity cost neglect. (2025). FunBlocks. https://www.funblocks.net/thinking-matters/classic-mental-models/opportunity-cost-neglect

...ıly 13). The cost of context switching (and
...*Work Life by Atlassian.* https://www.
...g/loom/cost-of-context-switching

Pejtersen, J. H., Feveile, H., Christensen, K. B., & Burr, H. (2011). Sickness absence associated with shared and open-plan offices: A national cross-sectional questionnaire survey. *Scandinavian Journal of Work, Environment & Health, 37*(5), 376–382. https://doi.org/10.5271/sjweh.3167

Priya, N. (2023). *Cortisol & your brain: How stress affects cognitive functions.* Click2pro. https://click2pro.com/blog/cortisol-affects-cognitive-functions

Ripla, A. (2025, October 6). *The psychology of crisis addiction in corporate leaders: Understanding the psychological drivers behind perpetual crisis management.* LinkedIn. https://www.linkedin.com/pulse/psychology-crisis-addiction-corporate-leaders-drivers-andre-xorfe/

Sander, E. (Libby) J., Marques, C., Birt, J., Stead, M., & Baumann, O. (2021). Open-plan office noise is stressful: Multimodal stress detection in a simulated work environment. *Journal of Management & Organization, 27*(6), 1–17. https://doi.org/10.1017/jmo.2021.17

Searle, E. (2023, January 31). *Flow state is the secret to a peak performance lifestyle, here's why...* LinkedIn. https://www.linkedin.com/pulse/flow-state-secret-peak-performance-lifestyle-heres-why-erin-searle/

Senge, P. M. (2006). *The fifth discipline: The art and practice of the learning organization.* Random House Business.

Sherbino, J., & Norman, G. (2021). Task switching, multitasking, and errors: A psychologic perspective on the impact of interruptions. *Annals of Emergency Medicine, 78*(3), 425–428. https://doi.org/10.101[illegible] annemergmed.2021.07.120

Shontell, A. (2013, February 4). *Sheryl Sandberg: "The most important document ever to come out of the Valley."* Business Insider. https://www.businessinsider.com/netflixs-management-and-culture-presentation-2013-2

Simon, H. (1973). *The organization of complex systems* (pp. 3–27). https://iiif.library.cmu.edu/file/Simon_box00066_fld05056_bdl0001_doc0001/Simon_box00066_fld05056_bdl0001_doc0001.pdf

Šmite, D., Moe, N. B., Floryan, M., Gonzalez-Huerta, J., Dorner, M., & Sablis, A. (2023). Decentralized decision-making and scaled autonomy at Spotify. *Journal of Systems and Software, 200,* Article 111649. https://doi.org/10.1016/j.jss.2023.111649

Smith, D. D. (2018, December 11). Brain-based tips for sharpening your focus. *Forbes.* https://www.forbes.com/councils/forbescoachescouncil/2018/12/11/brain-based-tips-for-sharpening-your-focus

Snowden, D., & Boone, M. (2007, November). *A leader's framework for decision making.* Harvard Business Review. https://hbr.org/2007/11/a-leaders-framework-for-decision-making

SYDLE. (2023, April 14). DORA metrics: What it is, its advantages, and how to adopt them in your company. *SYDLE.* https://www.sydle.com/blog/dora-metrics-643980428d074a29d2677746

Thomas, C. (2025, August 7). *The myth of the hero worker: Why businesses break when only a few hold it together.* LinkedIn. https://www.linkedin.com/pulse/myth-hero-worker-why-businesses-break-when-only-few-hold-corbelli-a7xlc/

Tomase, J. (2025). *Corporate AI not delivering yet.* LinkedIn. https://www.linkedin.com/news/story/corporate-ai-not-delivering-yet-6488140/

Tost, L. P., Gino, F., & Larrick, R. P. (2013). When power makes others speechless: The negative impact of leader power on team performance. *Academy of Management Journal, 56*(5), 1465–1486. https://doi.org/10.5465/amj.2011.0180

Warner, A., Davidovič, Š., Hidalgo, A., Beyer, B., Smith, K., & Duftler, M. (n.d.). *Canarying Releases, Google SRE Book.* Google. https://sre.google/workbook/canarying-releases/

Wei, M. (2024). 6 tips on being more assertive to reduce stress and anxiety. *Psychology Today.* https://www.psychologytoday.com/nz/blog/urban-survival/202409/6-tips-on-how-to-be-assertive-to-reduce-stress-and-anxiety

Wong, M. (2023, May 23). *What are DORA metrics?* Planview. https://www.planview.com/resources/articles/what-are-dora-metrics/

Wokittel, J. (2023, January 29). *What Conway's Law means for your organization.* Medium. https://jwokittel.medium.com/what-conways-law-means-for-your-organization-505032221057

Zimmerman, M. A. (1990). Toward a theory of learned hopefulness: A structural model analysis of participation and empowerment. *Journal of Research in Personality,* 24(1), 71–86. https://doi.org/10.1016/0092-6566(90)90007-s

The Refactor Ripple Effect

Help the Right CTO Find This Book

Most CTOs are stuck in reactive mode.
Few are intentionally designing their role.

If Refactor helped you think more clearly, reclaim autonomy, or shift from firefighting to leverage, please leave a review.

That's how better leadership spreads.

Scan the QR code or visit:
https://ctorefactor.com/review

www.ingramcontent.com/pod-product-compliance
Lightning Source LLC
Chambersburg PA
CBHW031324230926
51848CB00001B/74
9781967830046